Trieste Notes in Physics

Ivan T. Todorov

Conformal Description of Spinning Particles

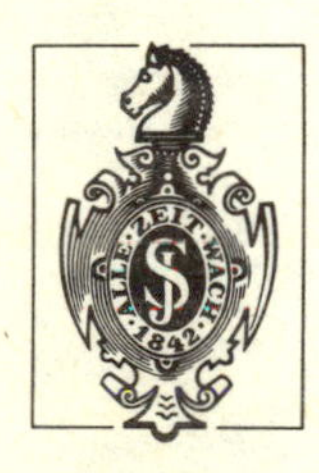

Springer-Verlag Berlin Heidelberg New York
London Paris Tokyo

Ivan T. Todorov

Scuola Internazionale Superiore di Studi Avanzati
I-34014 Trieste, Italy
and
Institute for Nuclear Research and Nuclear Energy
Bulgarian Academy of Sciences
Sofia 1184, Bulgaria

ISBN 3-540-16890-7 Springer-Verlag Berlin Heidelberg New York
ISBN 0-387-16890-7 Springer-Verlag New York Berlin Heidelberg

Library of Congress Cataloging-in-Publication Data. Todorov, Ivan T. Conformal description of spinning particles. (Trieste notes in physics) 1. Twistor theory. 2. Particles (Nuclear theory) 3. Conformal invariants. 4. Representation of groups. 5. Spinor analysis. I. Title. II. Series. QC173.75.T85T64 1986 539.7'21 86-15641

Offset printing and bookbinding: Druckhaus Beltz, D-6944 Hemsbach/Bergstr.
2153/3150-543210

Preface

These notes arose from a series of lectures first presented at the Scuola Internazionale Superiore di Studi Avanzati and the International Centre for Theoretical Physics in Trieste in July 1980 and then, in an extended form, at the Universities of Sofia (1980-81) and Bielefeld (1981). Their objective has been two-fold. First, to introduce theorists with some background in group representations to the notion of twistors with an emphasis on their conformal properties; a short guide to the literature on the subject is designed to compensate in part for the imcompleteness and the one-sidedness of our review. Secondly, we present a systematic study of positive energy conformal orbits in terms of twistor flag manifolds. They are interpreted as classical phase spaces of "conformal particles"; a characteristic property of such particles is the dilation invariance of their mass spectrum which, therefore, consists either of the point zero or of the infinite interval

$$0 < -p^2 \equiv p_0^2 - \mathbf{p}^2 < \infty \quad .$$

The detailed table of contents should give a thorough idea of the material covered in the text.

The present notes would have hardly been written without the encouragement and the support of Professor Paolo Budinich, whose enthusiasm concerning conformal semispinors (a synonym of twistors) — viewed in the spirit of Eli Cartan — is having a stimulating influence. In particular, our discussions on the role of the manifestly conformal covariant 8-component formalism in particle physics have induced me to include an appendix on the Clifford algebra of $O(6,\mathbb{C})$ and its "bitwistor" representation, where the traditional point of view on conformal Dirac spinors is restated.

I have profited from discussions both with the masters in the field and with my younger collaborators. I am particularly grateful to Dr. P. Nikolov whose help in the final stage of the work on these notes — both technical and concerning the substance of the text — has been instrumental in eliminating some mathematical inaccuracies and was crucial for completing the manuscript in time.

I would like to thank Professor P. Budinich for hospitality at the Scuola Internazionale Superiore di Studi Avanzati, Trieste and also Professor Abdus Salam, the International Atomic Energy Agency and UNESCO for hospitality at the International

Centre for Theoretical Physics, Trieste, where most of the work on these notes was done. I also benefitted from the hospitality of Professor Ph. Blanchard and Professor L. Streit at the University of Bielefeld where part of the notes were written and of Professor M. Vénéroni at the Institut de Physique Nucléaire in Orsay where the present revised version was prepared. In correcting some errors of the preprint version (in March 1984) before submitting the manuscript for the SISSA monograph series I am profiting from the very valuable advice and comments of Professor Walter Thirring.

Contents

Introduction

The conformal group — the group of angle preserving transformations of a pseudo-Riemannian manifold — seems to be playing a fundamental role in nature, a role which is yet to be fully understood and appreciated.

It first appeared in physics (as early as 1909) as the symmetry group of the free Maxwell equations. More general conformal invariant zero-mass field equations were studied in the 1930's. In the 1940's the conformal group was related to uniformly accelerated motion. In the 60's it was recognized as the group of causal automorphisms of space time. In the 70's the conformal symmetry was applied to (possible) critical behaviour in quantum field theory[1]. The twistors — the "quarks" (i.e. the lowest order faithful representations) of the conformal group — were introduced under this name in the late 60's and started gaining popularity in the last 3 or 4 years in relation to their successful application to the description of instantons. (See bibliographical note at the end of this introduction.)

The present notes are centred around the kinematical description (classical and quantum) of spinning particles.

Conventionally, both mass and spin are introduced as labels for the unitary irreducible representations (UIR's) of the Poincaré group. The notion of spin is, however, also a dilation invariant concept; it is, therefore, an invariant (and a label for the irreducible representations) of the 11-parameter group of Poincaré transformations and dilations which can be identified with the group Aut P of automorphisms of the Poincaré group (it is also called sometimes the Weyl group). As we shall see in the present notes the concept of spin is naturally associated with some coadjoint orbits of the conformal group SU(2,2) and with its corresponding UIR's. Indeed, as demonstrated by *Mack* [M1] every positive energy UIR of SU(2,2) (or of its universal covering) is either irreducible or splits into a finite number of UIR's of Aut P. The present notes contain a systematic study of the positive energy conformal orbits which give rise to the classical phase spaces of conformal

---

1 The work on this last topic is reviewed in [T5] where a historical survey and a comprehensive bibliography on the conformal group and its physical applications until 1978 is also given. (For more recent developments see [T6].)

particles. The role of the coadjoint orbits of a Lie group in both classical and quantum mechanics as well as in representation theory was first recognized by *Kirillov* (see Sect.15 of his monograph [K1]). The significance of the type of orbit in the explanation of symmetry and its breaking was demonstrated by the analysis of *Michel* and *Radicati* [M9] (in the case of a compact internal symmetry group). The analysis of conformal group orbits in terms of twistor flag manifolds (that is pursued here) was initiated by *Odziewicz* ([O1,2][L3]).

A Guide to the List of References

There are several different ways to approach the notion of twistors corresponding to their different applications. The richness and diversity of that notion are specially emphasized by *Penrose* (see, in particular, [P4-6] and [P11]). From this point of view the one-sidedness of our review (which mostly deals with conformal properties of twistors and with the twistor flag manifolds as phase spaces of "conformal particles") may be misleading. In order to compensate in part for this incompleteness, we have attempted to provide a rather comprehensive list of references on the subject.

The origin of twistor theory is usually traced back to studies of the conformal group and 0-mass particle wave equations (one should probably start with the early work of *Dirac* [D1] and then go to the papers [M11,12],[R1] and [H4] cited by Penrose). Penrose himself, the chief protagonist of the subject, has presumably come to the notion of twistors via his study of the conformal treatment of infinity [P0]. Only the easiest part of his and his collaborator's work [P1-14], [A3], [D4], [E1-4], [H1,2,7,8], [K2,3], [L1], [N1,2], [S2], [T1], [W1,10,11] dealing with flat space twistors is reviewed in the subsequent pages; concerning possible applications of twistors to general relativity the reader is referred to the original publications.

Although the main emphasis of Penrose and co-workers has been on Minkowski space twistors, the first really convincing application of the theory came in finding classical solutions of *Euclidean* Yang-Mills equations [W2], [A7,8,10-12]. The ideas of integral geometry, involved in this work, were also implemented in dealing with (simpler) linear problems (including the Maxwell equations) in [P8], [G1,2], [W5-7]. Further applications of complex Minkowski space twistors to the study of both self-dual and more general solutions of the Yang-Mills equations can be found in [B3], [I1], [M6], [W3,9]. For an extension of this technique to the study of gravitational instantons see [H5,6], [T2]. For application of quaternionic analyticity to that subject see [G4] (cf. also the critical review [M5]).

The twistor particle physics programme is still in a rather speculative stage (for a recent review and a subsequent development see [H7,8]). About the twistor description of massive particles and fields see [P12], [E1]. We already mentioned the study [O1,2], [L3] of phase spaces for conformal particles in terms of twistor flag manifolds.

Other paths leading to conformal semispinors (= twistors) are followed in [A1,2], [R2] and in [B6-8], where also different interpretations and applications are suggested.

For a background on the conformal group we refer to [D1], [K5], [L2], [M1-4] as well as to the book [T5] which also contains an extensive bibliography. The treatment of quantum 0-mass particles (see Chap.4) follows [M3] (see also [S4]).

Our references to papers concerned with zero mass particles (other than those involving twistors) are not at all systematic. The problem of existence of a position operator for 0-mass particles is tackled from different points of view in [W8],[A4], and [B2]. [W4] is a standard reference concerning Feynman rules for massless particles. A new approach to 0-mass particles in de Sitter space is being developed in [A6], [F3] (and references therein).

1. The Conformal Group of a Conformally Flat Space Time and Its Twistor Representations

1.1 Conformal Classes of Pseudo-Riemannian Metrics

Let (M,g) and ('M,'g) be two D-dimensional real pseudo-Riemannian manifolds, whose metric tensors g and 'g have the same signature. A diffeomorphism φ: U $\to$'U of a neighbourhood $U \subset M$ onto $'U \subset 'M$ is said to be *conformal* if the corresponding tangent map φ^T preserves the angles. In a slightly more technical language this means that for given coordinates X^μ and $'X^\mu$ on U and 'U we have

$$\frac{\partial\varphi^\kappa(x)}{\partial x^\mu}\frac{\partial\varphi^\lambda(x)}{\partial x^\nu}\, 'g_{\kappa\lambda}(\varphi(x)) = \Omega^2(x) g_{\mu\nu}(x) \qquad (\Omega > 0) \quad . \tag{1.1}$$

Thus conformal mappings are a generalization of *isometries* (i.e. metric preserving diffeomorphisms, corresponding to $\Omega = 1$). The pseudo-Riemannian manifolds (M,g) and ('M,'g) are said to be *conformally equivalent* (or simply *conformal*) if there is a conformal diffeomorphism φ: M $\to$'M and *locally conformal* if for any point of the one manifold there is a neighbourhood (containing that point) conformal to some neighbourhood in the other. (This definition includes situations in which the spaces M and 'M are not globally diffeomorphic; in particular, if M is non-compact, 'M could be a compactification of M.) A pseudo-Riemannian manifold M is said to be *conformally flat* if it is locally conformal to a (flat) pseudo-Euclidean space (of the same signature). In the set of all metric tensors on a given manifold M the relation $g \sim 'g$, if there exists a function $\Omega > 0$ such that $'g = \Omega^2 g$, is an equivalence relation and the equivalence classes are called *conformal classes of metrics.*

Remark. One can (see, e.g., *Manin* [M6]) define a global notion of a conformal structure on a differentiable manifold (corresponding locally to a conformal class of pseudo-Riemannian metrics).

Let F be the space of all frames on a real D-dimensional manifold M. (It is isomorphic to a principal fibre bundle with structure group GL(D,$\mathbb{R}$) — see [K3].) A pseudo-Riemannian metric can be defined as a section (in an associated bundle) of non-degenerate bilinear forms on TM of a given signature (p,q) (p + q = D) (with fibres isomorphic to the homogeneous space GL(D,$\mathbb{R}$)/O(p,q). A *conformal structure* is defined as a section in the corresponding projective fibre bundle whose elements are rays (through the origin) in the cone of all metrics of a given signature.

We shall not use this rather abstract notion in the present notes.

We say that a coordinate transformation $x \to {}'x(x)$ of a neighbourhood U of a pseudo-Riemannian manifold M is a *conformal symmetry* if

$$g_{\kappa\lambda}({}'x)\partial_\mu {}'x^\kappa \partial_\nu {}'x^\lambda = \Omega^2(x) g_{\mu\nu}(x) \partial_\alpha \equiv \frac{\partial}{\partial x_\alpha} \quad . \tag{1.2}$$

In other words, a conformal mapping [of type (1.1)] is a conformal symmetry, if $M = {}'M$ and ${}'g_{\mu\nu}({}'x) = g_{\mu\nu}({}'x)$, ${}'x = \varphi(x)$.

Not every pseudo-Riemannian manifold admits a non-trivial group of conformal symmetry transformations. In order to derive a condition for the existence of a one-parameter conformal symmetry group, we consider an infinitesimal transformation of type (1.2):

$${}'x^\mu(x) = x^\mu + \varepsilon K^\mu(x) + O(\varepsilon^2) \ , \quad \Omega(x) = 1 + \varepsilon f(x) + O(\varepsilon^2) \quad . \tag{1.3}$$

Inserting (1.3) into (1.1) and keeping the first-order terms in ε we find the *conformal Cartan-Killing equation*

$$\left[g_{\mu\lambda}(x)\partial_\nu + g_{\lambda\nu}(x)\partial_\mu\right]K^\lambda(x) + K^\lambda(x)\partial_\lambda g_{\mu\nu}(x) = 2fg_{\mu\nu} \tag{1.4a}$$

$$\nabla_\mu K_\nu + \nabla_\nu K_\mu = 2fg_{\mu\nu} \qquad \left(f = \frac{1}{D}\nabla_\mu K^\mu\right) \tag{1.4b}$$

where ∇_μ are the *covariant derivatives*

$$\nabla_\mu K_\nu = \partial_\mu K_\nu - \Gamma^\lambda_{\mu\nu} K_\lambda \quad (\nabla_\mu K^\nu = \partial_\mu K^\nu + \Gamma^\nu_{\lambda\mu} K^\lambda) \quad , \tag{1.5a}$$

$\Gamma^\lambda_{\mu\nu}$ being the *Christoffel symbols*:

$$\Gamma^\lambda_{\mu\nu} = \frac{1}{2} g^{\lambda\sigma}(\partial_\mu g_{\sigma\nu} + \partial_\nu g_{\mu\sigma} - \partial_\sigma g_{\mu\nu}) \quad . \tag{1.5b}$$

(In the case of an isometry, $f = 0$.) Spaces that are conformal to one another have isomorphic conformal symmetry groups.

Remark. We note that the form (1.4a) of the conformal Cartan-Killing equation is more general than the form (1.4b) that uses covariant derivatives, since it does not assume the nondegeneracy of the metric tensor [and hence the existence of the inverse tensor $g^{\lambda\mu}$ which enters the definition (1.5b) of a Riemannian affine connection $\Gamma^\lambda_{\mu\nu}$]. In fact, the relevant concept in the study of a symmetry is the concept of a Lie derivative. The Lie derivative of an arbitrary tensor valued function $T^{\lambda_1 \ldots \lambda_p}_{\mu_1 \ldots \mu_q}(x)$ along a vector field $K = K^\mu(x)\partial_\mu$ is defined as follows. Let ${}'x(t,x)$ be the (local) solution of the equation $(d/dt){}'x = K{}'x$ with initial condition ${}'x(0,x) = x$, which is given (formally) by ${}'x(t,x) = e^{tK}x$ and let $x(t,{}'x)$ be the inverse functions: $x(t,'x) = e^{-t'K}\,'x$, where $'K = K^{\mu}('x)(\partial/\partial' x^{\mu})$. Set

$$'T^{\lambda_1\ldots\lambda_p}{}_{\mu_1\ldots\mu_q} = \partial_{\lambda_1'} x^{\lambda_1} \ldots \partial_{\lambda_p'} x^{\lambda_p} T^{\lambda_1'\ldots\lambda_p'}{}_{\mu_1'\ldots\mu_q'} \partial_{\mu_1}{}' x^{\mu_1'} \ldots \partial_{\mu_q}{}' x^{\mu_q'}$$

$$\left(\partial_{\lambda'} = \frac{\partial}{\partial' x^{\lambda'}}\right) .$$

The *Lie derivative* L_K of T is defined by

$$L_K T^{\lambda_1\ldots\lambda_p}{}_{\mu_1\ldots\mu_q}(x) \quad \frac{d}{dt}\, 'T^{\lambda_1\ldots\lambda_p}{}_{\mu_1\ldots\mu_q}['x(t,x)]\Big|_{t=0} .$$

In particular, for a (contravariant) vector field

$$F = F^{\mu}\partial_{\mu}$$

we find

$$L_K F = [K,F] \quad \text{or} \quad L_K F^{\lambda} = K^{\mu}\partial_{\mu}F^{\lambda} - F^{\mu}\partial_{\mu}K^{\lambda}$$

while for a 1-form (covariant vector field) $\theta = \theta_{\mu} dx^{\mu}$

$$L_K\theta = \left(K^{\lambda}\partial_{\lambda}\theta_{\mu} + \theta_{\lambda}\partial_{\mu}K^{\lambda}\right)dx^{\mu} .$$

The explicit action of the Lie derivative on higher rank tensors is obtained from here by taking appropriate direct sums. In particular, the conformal Cartan-Killing equation (1.4a) can be written in the form

$$L_K g_{\mu\nu} = 2f g_{\mu\nu} . \tag{1.4c}$$

Example. Let M be the 3-sphere S^3 with the natural O(4)-invariant metric form. In terms of the spherical coordinates ρ,θ,φ (for which S^3 can be given by the parametric equations

$$\xi_1 = \sin\rho\,\sin\theta\,\cos\varphi, \quad \xi_2 = \sin\rho\,\sin\theta\,\sin\varphi, \quad \xi_3 = \sin\rho\,\cos\theta, \quad \xi_4 = \cos\rho)$$

the metric form is

$$ds^2 = d\rho^2 + \sin^2\rho(d\theta^2 + \sin^2\theta d\varphi^2) . \tag{1.6}$$

Exercise 1.1. Check that the SO(4) rotations of the unit vector ξ(in $\mathbb{R}^4$) are isometries [so that the corresponding Killing vectors satisfy (1.4) with $f = 0$]. Verify that the transformation $\xi \to '\xi$ where

$$'\boldsymbol{\xi}(\alpha,\xi) = \frac{\boldsymbol{\xi}}{\mathrm{ch}\alpha - \xi_4 \mathrm{sh}\alpha} , \quad '\xi_4(\alpha,\xi) = \frac{\xi_4\mathrm{ch}\alpha - \mathrm{sh}\alpha}{\mathrm{ch}\alpha - \xi_4\mathrm{sh}\alpha} \quad [\boldsymbol{\xi} = (\xi_1,\xi_2,\xi_3)] ,$$

is a conformal map of the unit sphere $S^3 = \{\xi ; \boldsymbol{\xi}^2 + \xi_4^2 = 1\}$ onto itself. (*Hint*: establish first that the corresponding Killing vector is

$$K \frac{\partial}{\partial \boldsymbol{\xi}} = \xi_4 \boldsymbol{\xi} \frac{\partial}{\partial \boldsymbol{\xi}} = \sin\rho \frac{\delta}{\delta\rho}$$

and verify the equation

$$(g_{\mu\lambda}\partial_\nu + g_{\lambda\nu}\partial_\mu)K^\lambda + K^\lambda \partial_\lambda g_{\mu\nu} = 2 \cos\rho g_{\mu\nu}$$

where

$\mu,\nu,\lambda = \rho,\theta,\varphi$, $g_{\mu\nu}$ is diagonal , $g_{\rho\rho} = 1$, $g_{\theta\theta} = \sin^2\rho$, $g_{\varphi\varphi} = \sin^2\rho\cos^2\theta$.)

The sphere S^3 also provides an example of a conformally flat Riemannian space. Indeed, the Euclidean space $\mathbb{R}^3$ can be inbedded in S^3 via the stereographic projection[1]

$$\mathbf{x} \to \boldsymbol{\xi} = \frac{2\mathbf{x}}{1 + \mathbf{x}^2} , \quad \xi_4 = \frac{1 - \mathbf{x}^2}{1 + \mathbf{x}^2}$$

$$\left(\mathbf{x} = \frac{\boldsymbol{\xi}}{1 + \xi_4} , \quad \mathbf{x}^2 = \frac{1 - \xi_4}{1 + \xi_4} , \quad \mathbf{x} \to \infty \Leftrightarrow \xi_4 \to -1\right) .$$

In terms of the x's the Riemannian metric (1.6) assumes the form

$$ds^2 = d\boldsymbol{\xi}^2 + d\xi_4^2 = \left(\frac{2}{1 + \mathbf{x}^2}\right)^2 d\mathbf{x}^2 . \tag{1.6'}$$

We see that the conformal factor $\Omega(x) = (2/1 + \mathbf{x}^2)$ does indeed go to zero for $\mathbf{x} \to \infty$.

This example is also instructive from the following point of view.

Given a (pseudo-) Riemannian manifold, it is not straightfoward to decide whether or not it is conformally flat according to the above definition, since an equality of the type $g_{\mu\nu}(x) = \Omega^2(x)\eta_{\mu\nu}$ is coordinate dependent. This point is amply illustrated by the above example. We have used three sets of local coordinates on S^3: (ρ,θ,ℓ), (ξ_1,ξ_2,ξ_3) and (x_1,x_2,x_3); only the third set exhibits the conformal flatness of the sphere. The question arises: is there an invariant characteristic of conformal flatness? A constructive affirmative answer to this question is given in Sect.1.2 below.

The conformal group is maximal for conformally flat spaces for which (locally)

$$g_{\mu\nu}(x) = \Omega^2(x)\eta_{\mu\nu} \tag{1.7}$$

where $\eta_{\mu\nu}$ is a pseudo-Euclidean metric with the same signature as $g_{\mu\nu}$. Let us describe the conformal group in this case for various values of the dimension D. In order to fix the language we shall start with the case of a (positive definite) Riemannian metric $g_{\mu\nu}$ and Euclidean η ($\eta_{\mu\nu} = \delta_{\mu\nu}$).

---

1 The stereographic projection is probably the oldest conformal mapping for which we find trace in the history of science: it has already been known to Ptolemy by the year 150 A.D.

For $D=1$ every (local) diffeomorphism $x \to 'x = \varphi(x)$ of the real line is conformal, since $d'x^2 = \varphi'^2 dx^2$. For $D=2$ the conformal group is also infinite dimensional. To see that we introduce a complex coordinate $z = x^1 + ix^2$ ($\bar{z} = x^1 - ix^2$) for which the conformally flat metric is a multiple of $dz\, d\bar{z}$. Every holomorphic transformation $z \to 'z = \varphi(z)$ ($\delta\varphi/\delta\bar{z} = 0$) is conformal since $d'z\, d'\bar{z} = |\varphi'(z)|^2 dz\, d\bar{z}$. The same is true for antiholomorphic transformations $z \to \varphi(\bar{z})$ (they change the orientation of angles but preserve their absolute values).

For $D \geqslant 3$ the conformal group is a $\frac{1}{2}(D+1)(D+2)$ dimensional Lie group isomorphic to the pseudoorthogonal groups $O(D+1,1)$ (see *Levine* [L2], as well as [E5], [P15], [D3]).

A simple-minded way to write down all infinitesimal conformal transformations in this case is to look for second-degree polynomial (in x) solutions of the Cartan-Killing equation (1.4). The result is

$$K_\mu (= \eta_{\mu\nu} K^\nu) = a_\mu + \omega_{\mu\nu} x^\nu + \alpha x_\mu + 2(xc)x_\mu - x^2 c_\mu \quad . \tag{1.8}$$

This formula remains true for a pseudo-Euclidean space of an arbitrary signature (p,q). The conformal group in this case is

$$O(p+1, q+1) \quad (\text{for} \quad p+q = D \geqslant 3) \quad .$$

Exercise 1.2. Find the conformal group for the 2-dimensional Minkowski space with metric form $ds^2 = d\mathbf{x}^2 - dt^2$.

(*Answer*: a general proper conformal transformation is a diffeomorphism of the form $\bar{x} \to 'x$ with $'\mathbf{x} = \frac{1}{2}[\varphi(\mathbf{x}+t) + \varphi(\mathbf{x}-t)]'t = \frac{1}{2}[\varphi(\mathbf{x}+t) - \varphi(\mathbf{x}-t)]$ where φ is a smooth real valued function on $\mathbb{R}$.)

1.2 Connection and Curvature Forms – a Recapitulation. The Weyl Curvature Tensor

An invariant characteristic of a conformal class of pseudo-Riemannian spaces is given by the Weyl curvature tensor. For the reader's convenience we shall reproduce here the necessary definitions using the language of differential forms and vector fields.

Let M be a differentiable manifold. We start by introducing corresponding to each other (dual) bases of 1-forms θ^α (in the cotangent bundle T^*M) and of vector fields E_β (in the tangent bundle TM).

In local coordinates $\theta^\alpha = \theta^\alpha_\mu dx^\mu$, $E_\beta = E^\nu_\beta \partial_\nu [\partial_\nu = \partial/\partial x^\nu, \theta^\alpha_\mu = \theta^\alpha_\mu(x), E^\nu_\beta = E^\nu_\beta(x)]$ a 1-form can be defined as a linear functional on vector fields, if we postulate that $dx^\mu(\partial_\nu) = \delta^\mu_\nu$. The *bases* $\{\theta^\alpha\}$ and $\{E_\beta\}$ are called *dual* if

$$\theta^\alpha(E_\beta) = \delta^\alpha_\beta \quad (\text{i.e., if} \quad \theta^\alpha_\mu E^\mu_\beta = \delta^\alpha_\beta) \quad . \tag{1.9}$$

Exercise 1.3. Show that the bases

$$\theta^i = \varepsilon_{ijk}\xi_k d\xi_j + \xi_4 d\xi_i - \xi_i d\xi_4$$

and

$$E_i = \xi_{ijk}\xi_k \frac{\partial}{\partial\xi_j} + \xi_4 \frac{\partial}{\partial\xi_i} - \xi_i \frac{\partial}{\partial\xi_4}$$ ($i,j,k = 1,2,3$, ε_{ijk} is the Levi-Civita antisymmetric unit tensor)

are dual to each other on the sphere S^3 and, moreover,

$$ds^2 = \delta_{ij}\theta^i\theta^j = (\theta^1)^2 + (\theta^2)^2 + (\theta^3)^2 \quad .$$

Express θ^i and E_j in terms of the independent coordinates x_k of (1.6').

A basis of 1-forms θ^α for which the metric form is $ds^2 = \eta_{\alpha\beta}\theta^\alpha\theta^\beta$ where η is the (constant) pseudo-Euclidean metric tensor is called a tetrad basis. The choice $\theta^\mu = dx^\mu$ $[E_\nu = \partial_\nu \equiv (\partial/\partial x^\nu)]$ is called a *coordinate basis*. A coordinate basis is characterized by the properties $d\theta^\mu = 0 = [E_\mu, E_\nu]$ where d is the exterior derivative, satisfying $d^2 = 0$, $[U,V] = (U^\mu\partial_\mu V^\nu - V^\mu\partial_\mu U^\nu)\partial_\nu$ is the commutator of vector fields). A tetrad basis can only coincide with a coordinate basis on a flat (pseudo-Euclidean) space.

A general (mixed) *tensor field* of contravariant rank p and covariant rank q is defined as an expression of the type

$$T^{\alpha_1 \ldots \alpha_p}_{\beta_1 \ldots \beta_q}(x) \quad E_{\alpha_1} \otimes E_{\alpha_2} \otimes \ldots \otimes E_{\alpha_p} \otimes \theta^{\beta_1} \otimes \ldots \otimes \theta^{\beta_q} \quad .$$

The *metric tensor* is defined as a symmetric second-rank covariant tensor:

$$ds^2 = g_{\alpha\beta}\theta^\alpha\theta^\beta \left[\equiv g_{\alpha\beta}\,\frac{1}{2}\,(\theta^\alpha \otimes \theta^\beta + \theta^\beta \otimes \theta^\alpha)\right]$$

which is assumed to be non-degenerate [i.e., to have $\det(g_{\alpha\beta}) \neq 0$]. The covariant derivative ∇_U along a vector field U is a rank preserving map of the set of smooth tensor fields into itself such that:

(i) $\nabla_{fU+gV} = f\nabla_U + g\nabla_V$ for any choice of the scalar functions f and g and vector fields U and V;

(ii) ∇_U is a differentiation, i.e. $\nabla_U(S \otimes T) = \nabla_U S \otimes T + S \otimes \nabla_U T$ and $\nabla_U f = Uf (= U^\mu\partial_\mu f$ in a coordinate basis);

(iii) the metric form is a covariant constant: $\nabla_U(g_{\alpha\beta}\theta^\alpha\theta^\beta) = 0$ (this condition, characteristic for a pseudo-Riemannian connection, is sometimes called *soldering condition*);

(iv) *absence of torsion*: $\nabla_U V - \nabla_V U = [U,V]$.

The covariant derivative satisfying (i) and (ii) is defined for an arbitrary tensor if we know the connection coefficients $\Gamma^\gamma_{\alpha\beta}$ defined by $\nabla_{E_\beta} E_\alpha = \Gamma^\gamma_{\alpha\beta}E_\gamma$ (in particular, then $\nabla_{E_\alpha}\theta^\beta = -\Gamma^\beta_{\sigma\alpha}\theta^\sigma$).

In a coordinate basis $\Gamma^{\lambda}_{\mu\nu}$ is symmetric in the lower indices and is called the Christoffel symbol. Moreover, the soldering condition then allows to express Γ in terms of the metric tensor and its derivatives — see (1.5b).

The information about the connection coefficients $\Gamma^{\alpha}_{\beta\gamma}$ can be coded in the *connection 1-forms* $\omega^{\alpha}_{\beta} = \Gamma^{\alpha}_{\beta\gamma}\theta^{\delta}$. They satisfy the *Cartan structure equations*

$$d\theta^{\alpha} = -\omega^{\alpha}_{\beta} \wedge \theta^{\beta} \tag{1.10}$$

(where $\wedge$ denotes the /skewsymmetric/ exterior product).

The (tensor-valued) *curvature 2-form* Ω^{α}_{β} is then defined by

$$\Omega^{\alpha}_{\beta} = d\omega^{\alpha}_{\beta} + \omega^{\alpha}_{\gamma} \wedge \omega^{\gamma}_{\beta} = \frac{1}{2} R^{\alpha}_{\beta\gamma\delta}\theta^{\gamma} \wedge \theta^{\delta} \tag{1.11}$$

where $R^{\alpha}_{\beta\gamma\delta}$ is called the *Riemann curvature tensor.*

In local coordinates,

$$R^{\kappa}_{\lambda\mu\nu} = \Gamma^{\kappa}_{\lambda\nu,\mu} - \Gamma^{\kappa}_{\lambda\mu,\nu} + \Gamma^{\sigma}_{\lambda\nu}\Gamma^{\kappa}_{\sigma\mu} - \Gamma^{\sigma}_{\lambda\mu}\Gamma^{\kappa}_{\sigma\nu}$$

where

$$\Gamma^{\kappa}_{\lambda\mu,\nu} = \partial_{\nu}\Gamma^{\kappa}_{\lambda\mu}$$

[and $\Gamma^{\kappa}_{\lambda\mu}$ is given by (1.5b)]. An invariant characterization of a flat space time is given by the vanishing of the Riemann curvature.

The (2-form valued) tensor $\Omega_{\alpha\beta} = g_{\alpha\sigma}\Omega^{\sigma}_{\beta}$ is skewsymmetric: $\Omega_{\alpha\beta} + \Omega_{\beta\alpha} = 0$. (This follows from the skewsymmetry of $\omega_{\alpha\beta}$ in a tetrad basis in which $g_{\alpha\beta} = \eta_{\alpha\beta}$ and $d\eta_{\alpha\beta} = 0 = \omega_{\alpha\beta} + \omega_{\beta\alpha}$ and from the tensor character of Ω.) Differentiating both sides of (1.10) we find

$$\Omega^{\alpha}_{\beta} \wedge \theta^{\beta} = 0 \quad \text{or} \quad R^{\alpha}_{\beta\gamma\delta} + R^{\alpha}_{\delta\beta\gamma} + R^{\alpha}_{\gamma\delta\beta} = 0 \quad . \tag{1.12}$$

Exercise 1.4. Show that $R^{\alpha}_{\beta\gamma\delta}$ has $\frac{1}{12} D^2(D^2 - 1)$ independent components (on a D-dimensional manifold. Prove the Bianchi identities:

$$d\Omega^{\alpha}_{\beta} = \Omega^{\alpha}_{\gamma} \quad \omega^{\gamma}_{\beta} - \omega^{\alpha}_{\gamma} \wedge \Omega^{\gamma}_{\beta} \quad . \tag{1.13}$$

The Ricci tensor $R_{\alpha\beta}$ and the scalar curvature R are defined as traces of the Riemann tensor

$$R_{\alpha\beta} = R^{\gamma}_{\alpha\gamma\beta} \; , \quad R = R^{\alpha}_{\alpha} \quad . \tag{1.14}$$

The (conformal) Weyl curvature form C^{α}_{β} is defined (for $D \geqslant 3$) by

$$\Omega^{\alpha}_{\beta} = \frac{R}{(D-1)(D-2)} \theta^{\alpha}\theta_{\beta} - \frac{1}{D-2} (\theta^{\alpha}R_{\beta\gamma}\theta^{\gamma} - \theta_{\beta}R^{\alpha}_{1}\theta^{\gamma}) + C^{\alpha}_{\beta} \tag{1.15a}$$

(see Sect.4.1.24 of [TO]). It gives rise to the traceless Weyl curvature tensor

$$C^{\alpha}_{\beta\gamma\delta} = R^{\alpha}_{\beta\gamma\delta} + \frac{1}{D-2} (R_{\beta[\gamma}\delta^{\gamma}_{\delta]} + g_{\beta[\gamma}R^{\alpha}_{\delta]}) - \frac{R}{(D-1)(D-2)} g_{\beta[\gamma}\delta^{\alpha}_{\delta]} \; , \tag{1.15b}$$

where the square brackets stand for antisymmetrization. The following result justifies the name 'conformal' for the Weyl tensor.

Proposition. If two diffeomorphic pseudo-Riemannian spaces (of dimension $D \geqslant 3$) are conformal to each other, then they have the same Weyl tensor. In particular, a space M is conformally flat if $C^{\alpha}_{\beta\gamma\delta} = 0$.

Sketch of the proof. Let $\bar{g}_{\mu\nu}(x) = e^{2f(x)} g_{\mu\nu}(x)$; set

$$f,_{\lambda} = \partial_{\lambda} f, \; f_{\lambda\mu} = f,_{\lambda\mu} - f,_{\lambda} \; f,_{\mu} \; ; \quad \text{then}$$

$$\bar{R}^{\kappa}_{\lambda\mu\nu} = R^{\kappa}_{\lambda\mu\nu} + f_{\lambda[\mu}\delta^{\kappa}_{\nu]} + g_{\lambda[\mu}f^{\kappa}_{\nu]} + (\partial f)^2 g_{\lambda[\mu}\delta^{\kappa}_{\nu]} \quad ,$$

$$\bar{R}_{\mu\nu} = R_{\mu\nu} - (D-2)f_{\mu\nu} - g_{\mu\nu}[\Box f + (D-2)(\partial f)^2]$$

$$\bar{R} = R - 2(D-1)\left[\Box f + \left(\frac{D}{2} - 1\right)(\partial f)^2\right]$$

and hence

$$\bar{C}^{\kappa}_{\lambda\mu\nu} = C^{\kappa}_{\lambda\mu\nu} \quad .$$

Remark. For non-conformally flat manifolds there are examples of metrics which give the same Weyl tensor without being conformally equivalent (see [A4]).

1.3 Global Conformal Transformations in Compactified Minkowski Space. Conformal Invariant Local Causal Order on $\overline{M}$

The global conformal transformations in Minkowski space, corresponding to the Killing vectors (1.8), are composed of pseudo-Euclidean motions $x \to \Lambda x + a$ ($\Lambda = e^{\omega}$), dilations $x \to e^{\alpha} x$, and special conformal transformations

$$x^{\mu} \to {}'x^{\mu}(c,x) = \frac{x^{\mu} + x^2 c^{\mu}}{1 + 2cx + c^2 x^2} \quad , \qquad \mu = 0,1,2,3, \quad cx = \mathbf{cx} - c^0 x^0 \quad . \tag{1.16}$$

The conformal factor

$$\Omega = \Omega(x,c) = (1 + 2xc + x^2 c^2)^{-1} \tag{1.17}$$

in this latter case is singular on the cone $(x + c/c^2)^2$ or on the isotropic hyperplane $2xc + 1 = 0$ for $c^2 = 0$.[2] Thus the Minkowski space M is not a proper carrier space for global conformal transformations. Such transformations make sense, however, on the compactified space $\bar{M}$ defined (according to *Dirac* [D1]) as the set of straight lines through the origin on the cone[3]

---

2 The cone of singularities is characteristic for the pseudo-Euclidean case. If the metric is positive definite (i.e., if the space is Euclidean) then the singularity set degenerates into a point.

3 In the mathematical literature Q is often called the Klein quadric (see e.g., [M5]). Klein's realization corresponds to a different basis which will be displayed in Sect.2.2.

$$Q(=Q_M) = \left\{\xi \in \mathbb{R}^6 \; ; \quad (\xi,\xi) = \eta_{ab}\xi^a\xi^b \equiv \xi^\mu \eta_{\mu\nu}\xi^\nu + (\xi^5)^2 - (\xi^6)^2 = 0 \; , \quad \xi \neq 0\right\} \; . \tag{1.18}$$

In other words, $\bar{M}$ is the projective space

$$\bar{M} = PQ = Q/\mathbb{R}^* \quad (\mathbb{R}^* \text{ is the set of non-zero reals}) \; . \tag{1.19}$$

In order to identify M as a (dense) open submanifold of $\bar{M}$ we have to single out a *null cone at infinity* J (in Penrose notation). A standard choice is the intersection of PQ with the hyperplane $\xi^5 + \xi^6 = 0$. There is a one to one correspondence between M and $\bar{M} \setminus J$ given by

$$x^\mu = \frac{1}{\kappa}\xi^\mu \; , \quad \kappa \equiv \xi^5 + \xi^6 \neq 0 \; (\xi^6 - \xi^5 = -\xi_6 - \xi_5 = \kappa x^2) \; . \tag{1.20}$$

(Here and in what follows we are using dimensionless variables. If we wish to recover space coordinates of dimension of length we should substitute x and ξ by x/ℓ and ξ/ℓ where ℓ is a fixed length.)

Every conformal transformation in M is generated by a linear O(4,2)-transformation on $\mathbb{R}^6$ (which leaves the cone Q invariant). In particular, the linear transformation $\xi \to {}'\xi(a,\xi)$ that corresponds to x-space translation $x \to {}'x = x - a$ is

$${}'\xi^\mu = \xi^\mu - \kappa a^\mu \; , \quad {}'\kappa = \kappa (\equiv \xi^5 + \xi^6) \; , \quad {}'\xi^5 - {}'\xi^6 = \xi^5 - \xi^6 + 2a\xi - a^2\kappa \; . \tag{1.21}$$

Exercise 1.5. Find the Minkowski space counterpart of the 1-parameter group of rotations in the (0,6)-plane:

$$\xi \to {}'\xi(\alpha,\xi) = (\cos\alpha\,\xi^0 + \sin\alpha\,\xi^6, \; \xi^1, \xi^2, \xi^3, \xi^5, -\xi^0\sin\alpha + \xi^6\cos\alpha) \; .$$

[*Answer*: $${}'x^0 = \frac{2x^0\cos\alpha + (1 + x^2)\sin\alpha}{1 - x^2 + (1 + x^2)\cos\alpha - 2x^0\sin\alpha} \; ,$$

$${}'\mathbf{x} = \frac{2\mathbf{x}}{1 - x^2 + (1 + x^2)\cos\alpha - 2x^0\sin\alpha} \; .]$$

The involutive proper conformal transformation corresponding to rotation on π in the (0,6)-plane is the *Weyl reflection*

$$I_W : x \to x_W \; , \quad x_W^0 = \frac{x^0}{x^2} \; , \quad \mathbf{x}_W = -\frac{\mathbf{x}}{x^2} \; . \tag{1.22}$$

(A transformation, say I_W, is called *involutive*, if $I_W^2 = 1$.) Every conformal transformation can be obtained by a multiple application of a Weyl reflection and translations.

The group O(4,2) has four connected components: (1) the connected subgroup $SO_0(4,2)$ (the component of the identity); (2) the component of SO(4,2) containing the space-time inversion

$$I_{st}x = -x \ , \quad I_{st}(\xi^\mu,\xi^5,\xi^6) = (-\xi^\mu,\xi^5,\xi^6) \quad ;$$

(3) the component of $O(4,2)$ containing the conformal inversion

$$I_r x = \frac{x}{x^2} \ , \quad I_r(\xi^0,\boldsymbol{\xi}^5,\xi^6) = (\xi^0,\boldsymbol{\xi},-\xi_5,\xi_6) \tag{1.23}$$

(and hence also the space reflection); (4) the connected component of $O(4,2)$ containing the time reflection $I_t(x^0,\mathbf{x}) = (-x^0,\mathbf{x})$. The union of (1) and (3), i.e., the set of orthochronous conformal transformations will be denoted by $O\uparrow(4,2)$. The connected component of the identity $SO_0(4,2)$ of the conformal group has a 2-element centre isomorphic to $\mathbb{Z}_2$.

The non-trivial element of the centre, the reflection of all six coordinates $\xi \to -\xi$, acts trivially on points of Minkowski space as it is obvious from (1.20). Thus the proper conformal group of space time is the simple (factor) group $SO_0(4,2)/\mathbb{Z}_2$. That is also the conformal group of compactified Minkowski space $\bar{M}$ (1.19) which is diffeomorphic to

$$\bar{M} = (S^3 x S^1)/\mathbb{Z}_2 \equiv \left\{\xi;\xi_1^2 + \xi_2^2 + \xi_3^2 + \xi_5^2 = \xi_0^2 + \xi_6^2 = 1\right\}/\xi \simeq - \xi \quad . \tag{1.24}$$

The space $\bar{M}$ has a natural pseudo-Riemannian structure, obtained as a restriction of the $O(4,2)$ invariant (on $\mathbb{R}^6$) metric form

$$ds^2 = (d\xi,d\xi) = -(d\xi^0)^2 + d\boldsymbol{\xi}^2 + (d\xi^5)^2 - (d\xi^6)^2 \quad \text{on} \quad S^3 x S^1 \quad .$$

Parametrizing the unit circle $(\xi^0)^2 + (\xi^6)^2 = 1$ by

$$\xi^0 = \sin\tau \ , \ \xi^6 = \cos\tau \tag{1.25}$$

and using the spherical coordinates ρ,θ,φ on S^3 (see Example in Sect.1.1) we find

$$ds^2 = (d\xi,d\xi) = d\rho^2 + \sin^2\rho(d\theta^2 + \sin^2\theta d\varphi^2) - d\tau^2 \ . \tag{1.26}$$

On the submanifold $\bar{M}\setminus J$ (where

$$J = \{-\pi < \tau \leqslant \pi \quad , \quad 0 \leqslant \rho \leqslant \pi \quad , \quad 0 \leqslant \theta < \pi \quad , \quad 0 \leqslant \varphi < 2\pi \quad ;$$
$$\cos\rho + \cos\tau = 0\})$$

we can use the x's as local coordinates, setting, for

$$\cos\tau + \cos\rho > 0 \ , \quad \sin\tau = \Omega(x)x^0 \ , \quad \cos\tau = \frac{1}{2}\,\Omega(x)(1 + x^2) \quad ,$$
$$\cos\rho = \frac{1}{2}\,\Omega(x)(1 - x^2)$$
$$\sin\rho(\sin\theta\cos\varphi, \sin\theta\sin\varphi, \cos\theta) = \Omega(x)(x^1,x^2,x^3) \quad , \tag{1.27}$$

where

$$\Omega(x) = [(x^0)^2 + \frac{1}{4}(1 + x^2)^2]^{-\frac{1}{2}} \ (= \cos\tau + \cos\rho) \quad . \tag{1.28}$$

In these coordinates ds^2 assumes a manifestly conformally flat form:

$$ds^2 = \Omega^2 dx^2 \; [dx^2 = \eta_{\mu\nu} dx^\mu dx^\nu = \mathbf{dx}^2 - (dx^0)^2] \quad . \tag{1.29}$$

Both M and $\bar{M}$ allow a Poincaré and dilatation invariant local causal ordering, originating from the partial ordering of the tangent space vectors, dx (at a point x) into future and past pointing (an ordering only defined for timelike and lightlike vectors for which $dx^2 \leqslant 0$). The causal order on (M or) $\bar{M}$ is also invariant under (local) special conformal transformations. In order to verify it for the special conformal transformations on M [of type (1.16)], we note that

$$d'x^\mu(c,x) = \frac{\Lambda(x,c)^\mu_\nu dx^\nu}{1 - 2xc + x^2c^2} \quad ,$$

where

$$\Lambda^\mu_\nu = \delta^\mu_\nu + 2\,\frac{(x^\mu - x^2c^\mu)c_\nu - c^\mu x_\nu(1 - 2cx) - c^2x^\mu x_\nu}{1 - 2cx + c^2x^2} \in SO^\uparrow(3,1) \quad .$$

Hence, for c sufficiently small (so that $2xc - x^2c^2 < 1$), $d'x$ is obtained from dx by a dilation (with a positive factor) and a proper Lorentz transformation both of which preserve the causal order. The causal order on each tangent space $T\bar{M}$ is defined, on the other hand, by the inequality

$$d\tau > \sqrt{d\rho^2 + \sin^2\rho(d\theta^2 + \sin^2\theta d\varphi^2)} \quad . \tag{1.30}$$

Its $O^\uparrow(4,2)$ invariance is a consequence of the fact that the conformal factor which multiplies ds^2 (1.26) under a proper conformal transformation never vanishes. [It is sufficient to verify the invariance of (1.30) under the Weyl reflection (1.22) (which corresponds to the τ translation $\tau \to \tau + \pi$) and under translations (1.21) since these two types of transformations generate the entire conformal group.]

Note finally that the causal order defined above has a global nature on M but not on $\bar{M}$ since $\bar{M}$ contains closed timelike curves (like the circle $\xi_0^2 + \xi_6^2 = 1$). However, the universal covering space $\tilde{M} = \mathbb{R} \times S^3$ of $\bar{M}$ (that is the cylinder $-\infty < \tau < \infty$, $0 \leqslant \rho \leqslant \pi$, $0 \leqslant \theta \leqslant \pi$, $0 \leqslant \varphi < 2\pi$) has a conformally invariant global causal order. (For a more detailed discussion of this problem we refer to *Segal* [S1] as well as to Sect.1.3 of the book [T4].)

1.4 The Lie Algebra of the Conformal Group and Its Twistor Representations

The Lie algrebra $so(4,2)$ of the (connected) pseudo-orthogonal group $SO_0(4,2)$ is spanned by the "momenta" J_{ab} [that generate (pseudo-)rotations in the plane (a,b), $a,b = 0,1,2,3,5,6$], satisfying the standard commutation relations

$$\frac{1}{i}[J_{ab},J_{cd}] = \eta_{ac}J_{bd} - \eta_{ad}J_{bc} + \eta_{bd}J_{ac} - \eta_{bc}J_{ad}[\eta_{ab} = \mathrm{diag}(-+\ldots+-)] \quad . \tag{1.31}$$

[J_{ab} are the so-called "physical generators" of the conformal group. For unitary representations of $SO_0(4,2)$ they are hermitian operators, which can be interpreted as physical observables. They are related to the "mathematical generators" X_{ab} (corresponding to real structure constants) by $J_{ab} = iX_{ab}$.]

The Lorentz angular momentum is identified with $J_{\mu\nu}(\mu,\nu = 0,1,2,3)$. In order to single out the translation generators P_μ corresponding to the representation $[T(a)f](x) = f(x-a)$ in the homogeneous coordinates realization (1.21), we first note that for the representation of $O(4,2)$ acting as coordinate transformations on scalar functions on the cone Q (1.18) J_{ab} are represented by the orbital angular momentum operators

$$L_{ab} = i\left(\xi_b \frac{\partial}{\partial\xi^a} - \xi_a \frac{\partial}{\partial\xi^b}\right) \tag{1.32}$$

and compare with

$$(P_\mu f)(\xi) = i \frac{\partial}{\partial a^\mu} f[\xi(a,\xi)]\Big|_{a=0} = -i\left[\kappa \frac{\partial}{\partial\xi^\mu} - \xi_\mu\left(\frac{\partial}{\partial\xi^5} - \frac{\partial}{\partial\xi^6}\right)\right]f(\xi) \quad .$$

The result (extended to an arbitrary representation) reads

$$P_\mu = J_{\mu 6} - J_{\mu 5} \quad . \tag{1.33}$$

Similarly, the special conformal generators are given by

$$K_\mu = J_{\mu 6} + J_{\mu 5} \quad . \tag{1.34}$$

(The signs are chosen in such a way that for a physical representation $P^0 \geqslant 0$ and $K^0 \geqslant 0$.) In terms of $J_{\mu\nu}$, P_μ, K_ν and the dilation generator J_{65}, the commutation relations (1.31) assume the form

$$\frac{1}{i}[J_{\mu\nu},P_\lambda] = \eta_{\mu\lambda}P_\nu - \eta_{\nu\lambda}P_\mu \ , \quad \frac{1}{i}[J_{\mu\nu},K_\lambda] = \eta_{\mu\lambda}K_\nu - \eta_{\nu\lambda}K_\mu \ ,$$

$$\frac{1}{i}[J_{65},P_\lambda] = P_\lambda \ , \quad \frac{1}{i}[J_{65},K_\lambda] = -K_\lambda \ , \quad [P_\mu,P_\nu] = [K_\mu, K_\nu] = 0 = [J_{\mu\nu},J_{65}]$$

$$\frac{1}{i}[P_\mu,K_\nu] = 2(\eta_{\mu\nu}J_{65} - J_{\mu\nu}) \quad . \tag{1.35}$$

The lowest-order faithful representations of the Lie algebra $so(4,2)$ are 4-dimensional. (This is the first time when the fact that the space-time dimension is $D = 4$ plays a role.) There are two inequivalent 4-dimensional irreducible representations of $so(4,2)$, corresponding to the *twistors* and their *duals*. To describe them we consider two sets of 4×4 matrices β_a and $\check{\beta}_a$ satisfying

$$\beta_a\check{\beta}_b + \beta_b\check{\beta}_a = \check{\beta}_a\beta_b + \check{\beta}_b\beta_a = 2\eta_{ab} \quad (a,b = 0,1,2,3,5,6) \quad . \tag{1.36}$$

An example of such a pair of matrix 6-vectors is given by the Cartan basis realization

$$\beta_\mu = \check{\beta}_\mu = -i\begin{pmatrix} 0 & \sigma_\mu \\ \tilde{\sigma}_\mu & 0 \end{pmatrix} \qquad \mu = 0,1,2,3$$

$$\beta_5 = \check{\beta}_5 = i\beta_1\beta_2\beta_3\beta_0 = \begin{pmatrix} \sigma_0 & 0 \\ 0 & -\sigma_0 \end{pmatrix}, \quad \check{\beta}_6 = -\beta_6 = 1 \quad ; \tag{1.37a}$$

here $(\sigma_\mu) = (\sigma_0, \boldsymbol{\sigma})$, $(\tilde{\sigma}_\mu) = (\sigma_0, -\boldsymbol{\sigma})$ where $\boldsymbol{\sigma}$ are the Pauli matrices and $\sigma_0 = 1_2$:

$$\sigma_1 = \begin{pmatrix} 0 & 1 \\ 1 & 0 \end{pmatrix}, \quad \sigma_2 = \begin{pmatrix} 0 & -i \\ i & 0 \end{pmatrix}, \quad \sigma_3 = \begin{pmatrix} 1 & 0 \\ 0 & -1 \end{pmatrix}, \quad \sigma_0 = \begin{pmatrix} 1 & 0 \\ 0 & 1 \end{pmatrix} . \tag{1.37b}$$

In verifying (1.36) one uses the identity

$$\sigma_\mu\tilde{\sigma}_\nu + \sigma_\nu\tilde{\sigma}_\mu = \tilde{\sigma}_\mu\sigma_\nu + \tilde{\sigma}_\nu\sigma_\mu = -2\eta_{\mu\nu} . \tag{1.38}$$

The mathematical generators of the twistor and dual twistor realizations of so(4,2) are given by

$$\gamma_{ab} = \frac{1}{4}(\beta_b\check{\beta}_a - \beta_a\check{\beta}_b) , \quad \check{\gamma}_{ab} = \frac{1}{4}(\check{\beta}_b\beta_a - \check{\beta}_a\beta_b) \tag{1.39}$$

(the corresponding physical generators being $i\gamma_{ab}$ and $i\check{\gamma}_{ab}$).

Exercise 1.6. Derive the infinitesimal transformation properties for β_a and $\check{\beta}_a$,

$$\gamma_{ab}\beta_c - \beta_c\check{\gamma}_{ab} = \eta_{ac}\beta_b - \eta_{bc}\beta_a \tag{1.40a}$$

$$\gamma_{ab}\beta_c - \beta_c\check{\gamma}_{ab} = \eta_{ac}\check{\beta}_b - \eta_{nc}\check{\beta}_a \quad , \tag{1.40b}$$

using only the defining relations (1.36) and (1.39). Setting

$$V_{(\lambda)} = \exp(\lambda^{ab}\gamma_{ab})^{-2} , \quad \check{V}_{(\lambda)} = \exp(\lambda^{ab}\check{\gamma}_{ab})^{-2} \quad (\lambda^{ab} = -\lambda^{ba}) \tag{1.41a}$$

$$\Lambda_{(\lambda)} = e^{-\lambda}\left(= \delta^a_b - \lambda^a_b + \frac{1}{2}\lambda^a_c\lambda^c_b + \ldots , \lambda^a_b = \lambda^{ac}\eta_{cb}\right) \tag{1.41b}$$

deduce the global transformation laws

$$V_{(\lambda)}\beta_a\check{V}^{-1}_{(\lambda)} = \beta_b\Lambda^b_{(\lambda)a} , \quad \check{V}_{(\lambda)}\beta_a V^{-1}_{(\lambda)} = \check{\beta}_b\Lambda_{(\lambda)}$$

Exercise 1.7. Check the commutation relations

$$[\gamma_{ab}, \gamma_{cd}] = \eta_{ac}\gamma_{bd} - \eta_{ad}\gamma_{bc} + \eta_{bd}\gamma_{ac} - \eta_{bc}\gamma_{ad}$$

(and similarly for $\check{\gamma}_{ab}$).

Remark. The expression (1.41b) for $\Lambda(\lambda)$ is in accord with (1.41a) in the following sense: if we set

$$-\lambda^a_b = \frac{1}{2}\lambda^{cd}(X_{cd})^a_b \quad \text{where} \quad (X_{cd})^a_b = \delta^a_d \eta_{cb} - \delta^a_c \eta_{db}$$

then X_{ab} satisfy the same commutation relations as γ_{ab}.

We note that (1.41c) or, equivalently,

$$\Lambda^a_b = \frac{1}{4}\,\mathrm{tr}(\check{\beta}^a V \beta_b \check{V}^{-1}) = \frac{1}{4}\,\mathrm{tr}(\beta^a \check{V} \check{\beta}_b V^{-1}) \tag{1.41d}$$

$$a,b = 0,1,2,3,5,6 \quad ,$$

establishes an isomorphism between $SO_0(4,2)$ and $SU(2,2)/\mathbb{Z}_2$ [the pairs $(V,\check{V})$ and $(-V,-\check{V})$ giving rise to the same Λ]. There exists a hermitian matrix A with two positive and two negative eigenvalues such that

$$\check{\beta}^*_a A + A\beta_a = 0 \ (A = A^*) \quad \text{and hence} \quad \gamma^*_{ab}A + A\gamma_{ab} = 0 = \check{\gamma}^*_{ab}A + A\check{\gamma}_{ab} \quad . \tag{1.42}$$

In the realization (1.37) A, normalized by the condition $\det A = 1$, can be taken as $\pm\beta_4$ where

$$\beta_4 \equiv i\beta_0 = \begin{pmatrix} 0 & 1 \\ 1 & 0 \end{pmatrix} \quad (\beta_4^2 = 1) \quad . \tag{1.43}$$

(The equality $A = \pm\beta_4$ is characteristic for the basis in which $\beta^*_\mu = \beta_\mu$ for $\mu = 1,\ldots,4$.)

Thus the Lie algebra so(4,2) is isomorphic to the Lie algebra su(2,2) of the pseudounitary group.

A straightforward way to prove the inequivalence of the representations (γ_{ab}) and $(\check{\gamma}_{ab})$ of su(2,2) consists in verifying that the Casimir operator

$$C_3 = \frac{1}{3!}\,\varepsilon_{abcdef}\, J^{ab}J^{cd}J^{ef} \tag{1.44}$$

has different eigenvalues for the two representations.

Exercise 1.8. Show that

$$C_3(i\gamma_{ab}) = -i\beta^0\beta^1\beta^2\beta^3\beta^5\beta^6 = \beta_1\beta_2\beta_3\beta_4\beta_5\beta_6 = 1 \quad ,$$

$$C_3(i\check{\gamma}_{ab}) = \check{\beta}_1\check{\beta}_2\check{\beta}_3\check{\beta}_4\check{\beta}_5\check{\beta}_6 = -1 \quad .$$

Every other irreducible 4 dimensional representation of su(2,2) is equivalent to one of these two. In particular, the representation, generated by $(-{}^t\gamma_{ab})$ (where the superscript ^{t} to the left of a matrix stands for transposition), is equivalent to $(\check{\gamma}_{ab})$ and vice versa. To see this we observe that there exists an intertwining operator (4×4 matrix) B satisfying

$${}^t\beta_a B + B\beta_a = 0 = {}^tB\check{\beta}_a + {}^t\check{\beta}_a B \quad (\det B = 1) \quad . \tag{1.45}$$

Indeed, it follows from (1.40) and (1.45) that

$$B\gamma_{ab}B^{-1} = -{}^t\check{\gamma}_{ab} \quad . \tag{1.46}$$

Exercise 1.9. Verify that in the basis (1.37) $B = \beta_1\beta_3 = -\begin{pmatrix} \varepsilon & 0 \\ 0 & \varepsilon \end{pmatrix}$, $\varepsilon = \begin{pmatrix} 0 & 1 \\ -1 & 0 \end{pmatrix}$ satisfies (1.45). [On the other hand (1.49) expresses the equivalence of $-\gamma^*_{ab}$ (and $-\check{\gamma}^*_{ab}$) with γ_{ab} (respectively $\check{\gamma}_{ab}$).

The basis (1.37) is distinguished by its simple Lorentz properties; in particular, the representation $\gamma_{\mu\nu} = \check{\gamma}_{\mu\nu}$ of the Lorentz subalgebra $sl(2,\mathbb{C})$ is reduced .

Exercise 1.10. Find the generators P_μ and K_μ for the above representations in the basis (1.37).

$$\left[\textit{Answer:}\quad P_\mu(\gamma_{ab}) = \begin{pmatrix} 0 & 0 \\ \tilde{\sigma}_\mu & 0 \end{pmatrix} = -K_\mu(\check{\gamma}_{ab}) \;,\; K_\mu(\gamma_{ab}) = \begin{pmatrix} 0 & \sigma_\mu \\ 0 & 0 \end{pmatrix} = -\check{P}_\mu(\gamma_{ab}) \quad .\right]$$

Exercise 1.11. Show that the conformal inversion I_r (1.23) generates an automorphism of the Lie algebra which exchanges the places of P_μ and K_ν: $I_r(P_\mu) = K_\mu, I_r(K_\mu) = P_\mu$, $I_r(J_{\mu\nu}) = J_{\mu\nu}, I_r(J_{56}) = -J_{56}$.

2. Twistor Flag Manifolds and SU(2,2) Orbits

2.1 Seven Flag Manifolds in Twistor Space. Conformal Orbits in $F_1 = PT$

We have come to the notion of twistor space and its dual as representation spaces for the two inequivalent 4-dimensional representations of the Lie algebra su(2,2). More generally, the *twistor space* T is a 4-dimensional complex vector space ($T = \mathbb{C}^4$). It carries the natural (linear) action of the general linear group GL $(4,\mathbb{C})$. We shall be interested, in particular, in properties of twistors which remain invariant under the real form U(2,2) only briefly mentioning the real form[1] $U(1) \times Spin(5,1)$ of $GL(4,\mathbb{C})$.

In that case we shall speak about Minkowski (respectively, Euclidean) space twistors and shall use the notation $T_M (T_E)$ for the space T whenever we wish to stress that it is regarded as representation space for U(2,2) [or $U(1) \times Spin(5,1)$].

Since T can be viewed as the basic representation space for the conformal group of space-time it is not surprising that the space-time concepts [like the notion of a point (=event) or of a light ray] can be defined in terms of twistors. To this end we need to consider not only vectors but also various subspaces (lines, planes etc.) in T. This leads us to the notion of a (twistor) flag manifold.

A *flag manifold* in $\mathbb{C}^n$ is the set of all nested sequences $L_1 \subset \ldots \subset L_r$ of subspaces of $\mathbb{C}^n$ of fixed dimensions:

$$F_{d_1,\ldots d_r} \; [= F_{d_1,\ldots,d_r}(\mathbb{C}^n)] = \{L_1 \subset \ldots \subset L_r \subset \mathbb{C}^n\} \; ;$$
$$\dim_{\mathbb{C}} L_1 = d_1 < \ldots < \dim_{\mathbb{C}} L_r = d_r < n \quad . \tag{2.1}$$

In particular,

$$F_k = G_{n,k} (= G_{n,k}(\mathbb{C})) \quad (F_1 = P(\mathbb{C}^n) = \mathbb{C}P_{n-1}) \tag{2.2}$$

is the (complex) Grassmann manifold of k-planes in $\mathbb{C}^n$ (which coincides for $k = 1$ with the (n-1)-dimensional complex projective space).

---

1 Spin (5,1) is the 2-fold (universal) covering of the (proper) pseudoorthogonal group $SO^{\uparrow}(5,1)$.

Exercise 2.1. Prove that $F_{d_1,\ldots,d_r}$ is a compact manifold of (complex) dimensions

$$\dim_{\mathbb{C}} F_{d_1,\ldots,d_r}(\mathbb{C}^n) = d_1(d_2 - d_1) + \ldots + d_{r-1}(d_r - d_{r-1}) + d_r(n - d_r) \quad . \tag{2.3}$$

{*Hint*: reduce the evaluation of $\dim_{\mathbb{C}} F$ to the repeated evaluation of the (complex) dimension of Grassmann manifolds; note that $G_{n,k}(\mathbb{C})$ can be regarded as a homogeneous space of the unitary group U(n):

$$G_{n,k} = U(n)/U(k) \times U(n-k) \quad , \tag{2.4}$$

so that

$$\dim_{\mathbb{C}} G_{n,k} = \frac{1}{2}\left[n^2 - k^2 - (n-k)^2\right] = k(n-k).\}$$

There are seven twistor flag manifolds which are displayed on the commutative diagram in Fig.2.1.

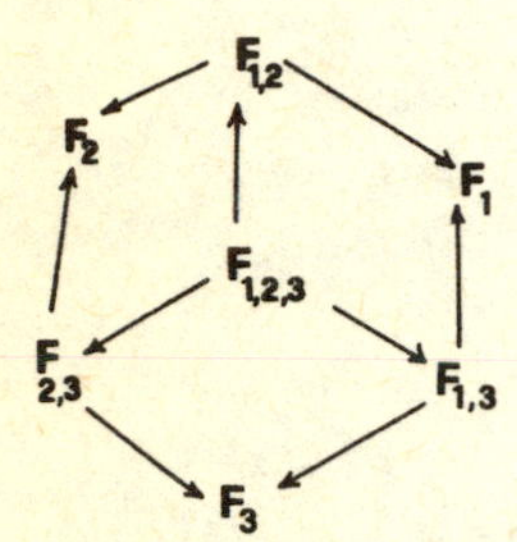

Fig.2.1. Diagram of all twistor flag manifolds. The arrows denote natural projections

The 3-dimensional manifold $F_3(T)$ of hyperplanes in T is isomorphic to the projective space $F_1(\tilde{T}) = P(\tilde{T})$ of dual twistors. If twistors and their duals are denoted by $\zeta = (\zeta^\alpha;\ \alpha = 1,2,3,4)\ (\in T)$ and $\tilde{\zeta} = (\tilde{\zeta}_\alpha)(\in \tilde{T})$ then the bilinear form

$$\tilde{\zeta}\zeta \equiv \tilde{\zeta}_\alpha \zeta^\alpha \left(\equiv \sum_{\alpha=1}^{4} \tilde{\zeta}_\alpha \zeta^\alpha\right) \tag{2.5}$$

is assumed to be GL(4,$\mathbb{C}$)-invariant. An element of $F_3(T)$, i.e. a hyperplane h through the origin in T, is given by the set of ζ satisfying an equation of the type $\tilde{\zeta}\zeta = 0$ for a fixed non-zero $\tilde{\zeta}$:

$$h = \{\zeta \in T;\ \tilde{\zeta}\zeta = 0\} \quad .$$

If λ is a non-vanishing complex number, then the substitution $\tilde{\zeta} \to \lambda\tilde{\zeta}$ does not alter the hyperplane h. Thus the "points" h of $F_3(T)$ are in one-to-one correspondence with the elements $[\tilde{\zeta}]$ of the projective space $P(\tilde{T}) = F_1(\tilde{T})$.
(Here $[\tilde{\zeta}]$ stands for the equivalence class of dual twistors of the form $\lambda\tilde{\zeta}$ for given $\tilde{\zeta}$ and λ running through $C^* = \mathbb{C} \setminus \{0\}$; $\tilde{\zeta}_\alpha$ play the role of homogeneous coordinates of a point in $P(\tilde{T})$.)

Note that there is no GL(4,$\mathbb{C}$) or U(1) × Spin(5,1)-invariant map from T onto $\tilde{T}$. (In other words, there is no invariant conjugation for complex or Euclidean space twistors.) For $\zeta \in T_M$, however, there exists a U(2,2)-invariant antilinear transformation

$$\zeta \to \tilde{\zeta} = \bar{\zeta}A \quad \text{where} \quad u^*A = Au^{-1} \quad \text{for} \quad u \in U(2,2) \tag{2.6}$$

[A is the hermitian matrix, defined by (1,41)]. (Of course, in the above identification of hyperplanes in T as equivalence classes $[\tilde{\zeta}] \in PT$, $\tilde{\zeta}$ is always regarded as an independent variable.)

Hence, the study of F_3 is reduced to the study of its "dual" F_1. The same type of argument establishes a ("duality") isomorphism between the flags $F_{1,2}$ and $F_{2,3}$. Thus we are left with the study of just five types of twistor flag manifolds: the 3-dimensional (complex) projective space F_1, the 4-dimensional Grassmann manifold F_2, the 5-dimensional flags $F_{1,2}$ and $F_{1,3}$ and the 6-dimensional manifold $F_{1,2,3}$ (in which all other flags are naturally imbedded).

The simplest flag manifold $F_1 = PT$ carries two open orbits F^+ and F^-, of the Minkowski space conformal group, SU(2,2), depending on the sign of $\tilde{\zeta}\zeta$ for $[\tilde{\zeta}] \in F_1$:

$$F^{\pm} = \{[\zeta] \in F_1 \quad ; \quad \pm\tilde{\zeta}\zeta > 0\} \quad .$$

[We note that although the U(2,2) invariant $\tilde{\zeta}\zeta$ is not a function of F_1 (only zero degree homogeneous functions on T can be regarded as functions of PT); its sign is one, since, for $\lambda \in C^*$, $\bar{\lambda}\tilde{\zeta}\lambda\zeta = |\lambda|^2\tilde{\zeta}\zeta$. Clearly this sign function is conformal invariant.]

Both F_1^+ and F_1^- are isomorphic to the homogeneous space

$$SU(2,2)/U(2,1) \quad .$$

The simplest way to see that is to choose a basis in which A is diagonal

$$A(=\beta_4) = \begin{pmatrix} 1 & 0 & 0 & 0 \\ 0 & 1 & 0 & 0 \\ 0 & 0 & -1 & 0 \\ 0 & 0 & 0 & -1 \end{pmatrix}$$

and to verify that the stability subgroup of each of the lines $[\zeta] = (\lambda,0,0,0)$ or $[\zeta] = [(0,0,0,\lambda)]$ is isomorphic to U(2,1). The two open orbits F_1^+ and F_1^- are separated (in F_1) by the degenerate 5-dimensional orbit

$$F_1^0 = \{[\zeta] \in F_1 \ ; \ \tilde{\zeta}\zeta = 0\} \quad .$$

A typical point on this orbit in the basis (1.37) is $[\zeta] = [(\lambda,0,0,0)]$. The stability subgroup of this point is a 10-parameter group $S_{10} \subset SU(2,2)$, whose description we leave to the reader.

2.2 Points of Compactified Space-Time as 2-Planes in Twistor Space

We shall show that points in F_2 — i.e., 2-planes in T — are in one-to-one correspondence with points of compactified complex Minkowski space $\overline{\mathbb{C}M}$.

The conformal compactification of complex Minkowski space $\mathbb{C}M$ is defined in the very same way as the compactification of real Minkowski space (see Sect.1.3). Setting

$$Q = Q_{\mathbb{C}} = \{Z \in \mathbb{C}^6 \; ; \; (Z,Z) = \eta_{ab} Z^a Z^b = 0 \; , \quad Z \neq 0\} \tag{2.7}$$

we define $\overline{\mathbb{C}M}$ as the corresponding (complex) projective manifold:

$$\overline{\mathbb{C}M} = P(Q_{\mathbb{C}}) = Q/_{\mathbb{C}^*} \quad (\mathbb{C}^* = \mathbb{C} \setminus \{0\}) \quad . \tag{2.8}$$

[In the complex domain we are not bound to use the real pseudo-Euclidean basis of Sect.1.3 in which the signature of the metric tensor η is (-+++,+-). For example the Euclidean basis, in which $(Z,Z) = \Sigma_{a=1}^{5} (Z^a)^2 - (Z^6)^2$, is also useful for certain purposes (cf. Appendix A).] The imbedding of CM into $\overline{CM}$ is given by the complex counterpart of (1.20):

$$z^\mu = \frac{1}{\kappa} Z^\mu \; , \quad \kappa = Z^5 + Z^6 \; (\neq 0), \; Z^6 - Z^5 = \kappa z^2 \quad . \tag{2.9}$$

Every $[Z] \in P(Q_{\mathbb{C}})$ defines a 2-plane in T (and a 2-plane in $\tilde{T}$) as the set of all $\zeta \in T$ (respectively all $\tilde{\zeta} \in T$) satisfying the equation

$$\check{\beta} Z \zeta (= \check{\beta}^a \eta_{ab} Z^b \zeta) = 0 \quad (\text{or} \quad \tilde{\zeta} \beta Z = 0) \quad . \tag{2.10}$$

Exercise 2.2. Let $\zeta \in T_M$ and $\tilde{\zeta} = \bar{\zeta} A$ as in (2.6). Show that each of the two equations $\check{\beta} Z \zeta = 0$ and $\tilde{\zeta} \beta \bar{Z} = 0$ follows from the other one. [In particular, for $Z \in Q_M$ the two equations (2.10) are equivalent.]

In order to see that (2.10) has two linearly independent solutions in T (as well as in $\tilde{T}$) so that it indeed defines a 2-plane in twistor space, it is sufficient to verify that

$$\text{rank} \; \check{\beta} Z = 2 (= \text{rank} \beta \bar{Z}) \qquad \text{for} \qquad Z \in Q_{\mathbb{C}} \quad . \tag{2.11}$$

To do this we first note that (2.10) is invariant under complex conformal transformations:

$$\zeta \to V(\lambda) \zeta \; (\tilde{\zeta} \to \tilde{\zeta} V^{-1}(\lambda)) \; Z \to \Lambda^{-1}(\lambda) Z \tag{2.12}$$

where

$$\Lambda^a_b = \frac{1}{4} \text{tr}(\check{\beta}^a V \beta_b \check{V}^{-1}) \tag{2.13a}$$

$$V(\lambda) = \exp\left(\frac{1}{2} \lambda^{ab} \gamma_{ab}\right) \; , \quad \check{V}(\lambda) = \exp\left(\frac{1}{2} \lambda^{ab} \check{\gamma}_{ab}\right) \quad (\lambda^{ab} = -\lambda^{ba}) \quad . \tag{2.13b}$$

Exercise 2.3. Verify the relations

$$V\beta Z\check{V}^{-1} = \beta\Lambda Z \;, \quad \check{V}\check{\beta}ZV^{-1} = \check{\beta}\Lambda Z \tag{2.14}$$

[where $V,\check{V}$ and Λ are given by (2.13)].

The invariance of (2.10) under the transformations (2.12) is a consequence of (2.14).

Remark. Equation (2.13a) provides a 2 to 1 homomorphism of $SL(4,\mathbb{C})$ onto $SO(6,\mathbb{C})$. Thus the simply connected group $SL(4,\mathbb{C})$ appears as the universal covering group of the complex conformal group. [Note that the real form $SU(2,2)$ of $SL(4,\mathbb{C})$ is not simply connected, its universal covering being infinite sheeted.]

Since $O_{\mathbb{C}}$ is a homogeneous space for $SO(6,\mathbb{C})$ and since rank βZ is obviously invariant under the complex conformal transformations $\check{\beta}Z \to V\check{\beta}ZV^{-1}$, it is sufficient to verify (2.11) for a special choice of Z [say, for $Z = (0,0,0,0,1,1)$ in the basis (1.37)] which is straightforward.

Exercise 2.4 Show that

$$(\beta Z)(\check{\beta}Z) = (\check{\beta}Z)(\beta Z) = Z^2 \quad .$$

Infer that $\det \beta Z = \det\check{\beta}Z = (Z^2)^2$.

Let $\underset{1}{\zeta}$, $\underset{2}{\zeta}$ be two linearly independent solutions of (2.10) (whose existence was just established). The 2-plane spanned by $\underset{1}{\zeta}$ and $\underset{2}{\zeta}$ is characterized by the *decomposable bivector*

$$f = \underset{1}{\zeta} \wedge \underset{2}{\zeta} \quad \text{with components} \quad f^{\alpha\beta} = \underset{1}{\zeta}^{\alpha}\underset{2}{\zeta}^{\beta} - \underset{1}{\zeta}^{\beta}\underset{2}{\zeta}^{\alpha} \quad \alpha,\beta = 1,\ldots,4 \quad . \tag{2.15}$$

Exercise 2.5. Demonstrate that a skewsymmetric second rank tensor is decomposable [i.e. has the form (2.15)] if

$$f\wedge f = (f^{\alpha\beta}f^{\gamma\delta} + f^{\gamma\alpha}f^{\beta\delta} + f^{\beta\gamma}f^{\alpha\delta}) = 0 = f^{12}f^{34} + f^{23}f^{14} + f^{31}f^{24} \quad . \tag{2.16}$$

It is clear that if $\underset{1}{\zeta}$ and $\underset{2}{\zeta}$ span a 2-plane in T and $S \in GL(2,\mathbb{C})$ then

$$'\underset{k}{\zeta} = \sum_{\ell=1}^{2} S_{k\ell}\, \underset{\ell}{\zeta} \quad (k = 1,2) \tag{2.17}$$

span the same 2-plane. It is easily verified that under such a change of basis the bivector f (2.15) is just multiplied by a non-zero complex number:

$$'f \equiv '\underset{1}{\zeta} \wedge '\underset{2}{\zeta} = (\det S)f \quad . \tag{2.18}$$

The *Klein quadric* (2.16) is isomorphic to the Dirac quadric Q (2.7). The correspondence $f \to Z$ is given by

$$Z_a = \frac{1}{4}\,\mathrm{tr}(fX_a) \quad , \qquad \text{where} \qquad X_a = {}^t B \gamma_a \tag{2.19}$$

[${}^tB(=-B^{-1})$ is the transpose to the matrix B defined by (1.45)].

Exercise 2.5 Prove the completeness relation

$$X_{a\alpha\beta} X^a_{\gamma\delta} = -2\varepsilon_{\alpha\beta\gamma\delta} \tag{2.20}$$

[*Hint*: Use the frame independence of (2.20) and check it in the basis (1.37) in which

$$X_1 = i\begin{pmatrix} 0 & -\sigma_3 \\ \sigma_3 & 0 \end{pmatrix}, \quad X_2 = \begin{pmatrix} 0 & 1 \\ -1 & 0 \end{pmatrix}, \quad X_3 = i\begin{pmatrix} 0 & \sigma_1 \\ -\sigma_1 & 0 \end{pmatrix},$$

$$X_4 = iX_0 = \begin{pmatrix} 0 & \varepsilon \\ \varepsilon & 0 \end{pmatrix}, \quad X_5 = \begin{pmatrix} \varepsilon & 0 \\ 0 & \varepsilon \end{pmatrix}, \quad X_6 = \begin{pmatrix} -\varepsilon & 0 \\ 0 & \varepsilon \end{pmatrix}, \quad \varepsilon = \begin{pmatrix} 0 & 1 \\ -1 & 0 \end{pmatrix}.]$$

Using (2.20), we see that the decomposability condition (2.16) is equivalent to the requirement $(Z,Z) = 0$ (2.7).

The inverse map $Z \to f$ is given by

$$f = {}^*XZ = \left(\frac{1}{2}\,\varepsilon^{\alpha\beta\gamma\delta} X^a_{\gamma\delta} Z_a\right) . \tag{2.21}$$

(To verify this use the identity $\frac{1}{2}\,\varepsilon^{\alpha\beta\gamma\delta}\varepsilon_{\alpha'\beta'\gamma\delta} = \delta^\alpha_{\alpha'}\delta^\beta_{\beta'} - \delta^\alpha_{\beta'}\delta^\beta_{\alpha'}$.)

The isomorphism between the two quadrics, thus established, leads (by going to the corresponding projective spaces) to the identification of the compactified complex Minkowski space $\overline{CM} = PQ$ and the Grassmann manifold $F_2 = G_{4,2}$.

Remark. The isomorphism between the quadrics (2.7) and (2.16) is given by a complex change of basis for both Euclidean and Minkowski space signature [where the real symmetry group of the metric tensor η_{ab} is O(5,1) and O(4,2) respectively]. It would be a real coordinate transformation for an O(3,3) symmetric η_{ab}.

2.3 An Alternative Realization of the Isomorphism $\overline{CM} \Leftrightarrow F_2$. SU(2,2) Orbits in the Grassmann Manifold

The introduction of twistors in the preceding sections and the display of their relation to the O(4,2) Clifford algebra in Appendix A is designed to allude to physicists' experience with Dirac spinors (and representation theory). There is a more direct way of displaying the correspondence $F_2 \Leftrightarrow \overline{CM}$ which should be more appealing to mathematicians (cf. [M5]).

We start with the familiar representation of points $z \in \mathbb{C}M$ in terms of 2×2 (complex) matrices

$$\underset{\sim}{z} = z^\mu \sigma_\mu = \begin{pmatrix} z^0+z^3 & z^1-iz^2 \\ z^1+iz^2 & z^0-z^3 \end{pmatrix} , \quad \tilde{z} = z^\mu \tilde{\sigma}_\mu = \begin{pmatrix} z^0-z^3 & -z^1+iz^2 \\ -z^1-iz^2 & z^0+z^3 \end{pmatrix} , \tag{2.22}$$

which satisfy $\underset{\sim}{z}\tilde{z} = -z^2 1 = \det z$, $\underset{\sim}{x}^* = \underset{\sim}{x}$, $\tilde{x} = \tilde{x}^*$ for x^μ real ($x \in M$). Each $\tilde{z}$ defines a map $\mathbb{C}^2 \to \mathbb{C}^2$: to every $\pi = (\pi^A)$ it makes correspond a $\bar{\chi} = (\bar{\chi}_{\dot{A}})$ satisfying

$$\tilde{z}_{\dot{A}B}\pi^B = i\bar{\chi}_{\dot{A}} \quad (i\underset{\sim}{z}\bar{\chi} = -z^2\pi) . \tag{2.23}$$

(The imaginary unit is introduced in the right-hand side for later convenience.)

This equation defines a 2-plane through the origin in the space $T = \mathbb{C}^4$ of twistors

$$\zeta = \begin{pmatrix} \pi \\ \bar{\chi} \end{pmatrix} . \tag{2.24}$$

Not all planes in T are of the form (2.23) (planes defined by $\pi = \underset{\sim}{p}\bar{\chi}$ for $\det \underset{\sim}{p} = 0$, cannot be set in that form). The manifold $F_2 = G_{4,2}$ of all 2-planes in T (through the origin) thus provides a non-trivial compactification of $\mathbb{C}M$.

We shall demonstrate that the correspondence $\overline{\mathbb{C}M} \Leftrightarrow F_2$ so obtained is the same as the one described in the preceding section.

Indeed, for $Z^\mu = \kappa z^\mu$ (2.9), using the basis (1.37) for $\check{\beta}_a$, we find

$$Z\check{\beta} = \kappa \begin{pmatrix} -z^2 & -i\underset{\sim}{z} \\ -i\tilde{z} & -1 \end{pmatrix} ; \tag{2.25}$$

hence, (2.23) implies $Z\check{\beta}\zeta = 0$ (2.10).

We now proceed to the description of the SU(2,2) orbits in $\overline{\mathbb{C}M}$.

The open conformal orbits in the space of 2-planes spanned by a pair of linearly independent twistors $\underset{1}{\zeta}, \underset{2}{\zeta}$, are characterized by the signature of the hermitian matrix

$$Z = \begin{pmatrix} \underset{1}{\tilde{\zeta}}\underset{1}{\zeta} & \underset{1}{\tilde{\zeta}}\underset{2}{\zeta} \\ \underset{2}{\tilde{\zeta}}\underset{1}{\zeta} & \underset{2}{\tilde{\zeta}}\underset{2}{\zeta} \end{pmatrix} . \tag{2.26}$$

Accordingly, there are three open orbits (of real dimension 8) corresponding to signatures (+,+), (+,-), and (-,-) (where the "+" and "-" stand for the signs of the eigenvalues of the matrix Z). It is not difficult to verify that the $(\pm,\pm)$ orbits correspond to the complex tube domains

$$T_\pm = \{z \in \mathbb{C}M \; ; \; \pm \operatorname{Im} z^0 > |\operatorname{Im} \underline{z}|\} \tag{2.27}$$

(which are primitive holomorphy domains for vacuum expectation values in a quantum field theory satisfying the spectrum condition). They are both isomorphic to the homogeneous space

$$F_2(\pm,\pm) = T_\pm \approx SO_0(4,2)/SO(4) \times SO(2) = SU(2,2)/S[U(2) \times U(2)] \tag{2.28}$$

the maximal compact subgroup of the conformal group being the stability group of the (complex) points $(z^k) = (\pm i, \underline{0})$ which are mapped by (2.21) into the bivectors

$$f_\pm = \begin{pmatrix} -\varepsilon & \mp\varepsilon \\ \mp\varepsilon & -\varepsilon \end{pmatrix} = \begin{pmatrix} 1 \\ 0 \\ \pm 1 \\ 0 \end{pmatrix} \wedge \begin{pmatrix} 0 \\ -1 \\ 0 \\ \mp 1 \end{pmatrix} \left[\varepsilon = \begin{pmatrix} 0 & 1 \\ 1- & 0 \end{pmatrix} \right] .$$

The (-,+) = (+,-) -orbit corresponds, on the other hand, to complex z with a space-like imaginary part. It is isomorphic to the homogeneous space

$$F_2(+,-) = SO_0(4,2)/SO_0(2,2) \times SO(2) \quad , \tag{2.29}$$

$SO_0(2,2) \times SO(2)$ being the stability subgroup of the point $z = (0,0,0,0)$ [or $Z = (0,0,0,i,1,0)$], which corresponds to the bivector

$$f_{(-,+)} = \begin{pmatrix} -\varepsilon & -\sigma_1 \\ \sigma_1 & \varepsilon \end{pmatrix} = \begin{pmatrix} -1 \\ 0 \\ 1 \\ 0 \end{pmatrix} \wedge \begin{pmatrix} 0 \\ 1 \\ 0 \\ 1 \end{pmatrix} .$$

The seven (real) dimensional orbits $(\pm,0)$ (corresponding to rank $Z = 1$) contain a pure imaginary point at infinity which can be taken as

$[Z] = [(\pm i,0,0,i,-1,1]$ (it corresponds to the bivector

$$f_{(\pm,0)} = \begin{pmatrix} 1 \\ 0 \\ \pm 1 \\ 0 \end{pmatrix} \wedge \begin{pmatrix} 0 \\ 1 \\ 0 \\ 0 \end{pmatrix} \quad).$$

Its stability subgroup is the 8-dimensional group generated by

$$J_{01} + J_{13},\ J_{02} + J_{23},\ J_{12},\ J_{15} + J_{16},\ J_{25} + J_{26},\ J_{03} + J_{56},\ J_{05} + J_{06},\ J_{35} + J_{36} .$$

Finally, the most degenerate 4-dimensional orbit, corresponding to $z = 0$, is isomorphic to the real compactified Minkowski space $\bar{M}$. The stability subgroup of a typical point, say $x = 0$ (or

$$[Z] = \left[\left(0,0,0,0,\tfrac{1}{2},\tfrac{1}{2}\right)\right], \quad f_{(0,0)} = \begin{pmatrix} 1 \\ 0 \\ 0 \\ 0 \end{pmatrix} \wedge \begin{pmatrix} 0 \\ -1 \\ 0 \\ 0 \end{pmatrix} \quad),$$

is the 11-parameter parabolic group S_{11}, compounded by Lorentz transformations, dilatations and special conformal transformations

$$\bar{M} = SO_0(4,2)/S_{11} \quad . \tag{2.30}$$

In the Euclidean quantum field theory it is also interesting to study the Spin(5,1) orbits of F_2.

Exercise 2.6. Show that $f = \underset{1}{\zeta} \wedge \underset{2}{\zeta}$ (2.15) defines a point of the compactified (real) Euclidean space S^4 if

$$\zeta_2 = \bar{B}\bar{\zeta}_1 \tag{2.31}$$

where B is the matrix defined by (1.45).

2.4 Higher Flag Manifolds

We turn to a brief description of the 5(complex)-dimensional flag manifolds $F_{1,2}$ and $F_{1,3}$.

The elements of $F_{1,2}$ are pairs, consisting of a 2-plane in T, represented by a point $[Z] \in PQ$ and a twistor line $[\zeta]$ belonging to that plane.

According to (2.10) the incidence condition $[\zeta] \subset [Z]$ is expressed analytically by the equation $Z\check{\beta}\zeta = 0$. Moreover, according to Exercise 2.2, if we are only interested in U(2,2)-invariant properties (if $T = T_M$), then (2.10) can be replaced by the equivalent equation $\tilde{\zeta}\beta\bar{Z} = 0$ where $\tilde{\zeta} = \bar{\zeta}A$ [is an U(2,2)-invariant relation].

The (open) SU(2,2) orbits of $F_{1,2}$ are again characterized by the signature of the form $\tilde{\zeta}\zeta$ in the corresponding subspaces. We shall see in Sect.3.3 below that the (-,-) orbit in $F_{1,2}(T_M)$ can be interpreted as the phase space of a (massive) spinning particle, the variable ζ carrying the spin degrees of freedom. The 2-planes with (+,-)-signature correspond to two orbits which depend on the signature (+ or -) of the distinguished 1-dimensional subspace of this plane. (Lower dimensional orbits are again obtained when the signature involves zeros.)

The flag $F_{1,3}$ can be identified with the set of (complex) light rays in $\overline{CM}$. Indeed, a light ray in $\overline{CM} = PQ$ is fixed by a pair of (different) points $[Z_{1,2}] \in PQ$ such that

$$(Z_1 - Z_2, Z_1 - Z_2) = -2(Z_1, Z_2) = 0 \quad . \tag{2.32}$$

Exercise 2.7. Show that if Z_1 and Z_2 correspond to finite points z_1 and z_2 in $\mathbb{C}M$ [according to (2.9)] then

$$-2(Z_1, Z_2) = \kappa_1\kappa_2(z_1 - z_2)^2 \quad . \tag{2.33}$$

In terms of the bivectors f_i (corresponding to Z_i) condition (2.32) can be written as

$$f_1 \wedge f_2 = 0 \quad . \tag{2.34}$$

Two linearly independent bivectors satisfying (2.34) correspond to 2-planes that intersect in an 1-dimensional subspace and whose union spans a 3-dimensional subspace of T. This is the 3-space with a singled out 1-dimensional line in it which gives an invariant identification of the light ray through Z_1 and Z_2.

Analytically $F_{1,3}$ is given by the set of pairs $\{['\tilde{\zeta}],[\zeta] \in PT \times \tilde{PT}$ such that $'\tilde{\zeta}\zeta = 0$. The SU(2,2) open orbits in $F_{1,3}$ are characterized by the pair (sign $'\tilde{\zeta}'\zeta$, sign $\tilde{\zeta}\zeta$).} Finally, a point in $F_{1,2,3}$ is parametrized by a triple $(['\zeta],[Z],[\zeta])$ such that

$$'\tilde{\zeta}(Z\beta) = 0 = (Z\check{\beta})\zeta(='\tilde{\zeta}\zeta) \quad . \tag{2.35}$$

A 12-dimensional SU(2,2) open orbit in $F_{1,2,3}$ has as a stability subgroup of a point 3-dimensional compact Abelian subgroup, generated by a Cartan subalgebra of SU(2,2).

Consider as an example the orbit $F^{+}_{1,2,3}$ for which the [Z]'s correspond to points in the future tube [i.e., such that the signature of the associated twistor 2-planes is (+,+)]. Then, as a consequence, $\tilde{\zeta}\zeta > 0 > '\tilde{\zeta}'\zeta$ [for ζ and $'\tilde{\zeta}$ satisfying (2.35)]. As a typical point on such an orbit we can take

$$Z = (i,0;0,1) \quad , \quad \zeta = \begin{pmatrix}1\\0\\1\\0\end{pmatrix} \quad , \quad '\tilde{\zeta} = (1,0,-1,0) \quad . \tag{2.36}$$

Exercise 2.8. Verify that the stability conformal subgroup of the triple $(['\tilde{\zeta}]\ [Z]\ [\zeta])$ (2.36) is generated by the (maximal) compact Cartan subalgebra

$$\lambda^{60}J_{60} + \lambda^{35}J_{35} + \lambda^{12}J_{12} \quad (\lambda^{60}, \lambda^{35}, \lambda^{12} \in \mathbb{R}) \quad . \tag{2.37}$$

3. Classical Phase Space of Conformal Spinning Particles

3.1 The Conformal Orbits F_1^+ and F_1^- as Phase Spaces of Negative and Positive Helicity 0-Mass Particles

The manifolds $F_1^{\pm}$ of Sect.2.1 admit a unique (up to a factor) SU(2,2)-invariant symplectic form

$$\omega_s = d\theta_s = \frac{2s}{\tilde{\zeta}\zeta i} d\tilde{\zeta}_\wedge \Pi_\zeta d\zeta \tag{3.1a}$$

$$= \frac{2s}{\tilde{\zeta}\zeta i}\left(d\zeta\wedge d\zeta - \frac{\zeta d\tilde{\zeta}\ \tilde{\zeta}d\zeta}{\tilde{\zeta}\zeta}\right) \quad \left(\Pi_\zeta = 1 - \frac{\zeta\otimes\tilde{\zeta}}{\tilde{\zeta}\zeta}\right) \tag{3.1b}$$

where

$$\theta_s = is\frac{\zeta d\tilde{\zeta} - \tilde{\zeta}d\zeta}{\tilde{\zeta}\zeta} \tag{3.2}$$

is a *real* conformal invariant 1-form on $F_1^{\pm}$.

The 2-form ω_s is the imaginary part of the SU(2,2)-invariant Kähler metric $(2s/\tilde{\zeta}\zeta)d\tilde{\zeta}_\alpha \otimes \Pi^\alpha_\beta d\zeta^\beta$ on $F_1^{\pm}$, which is called the *Fubini-Study metric*.

The standard evaluation of Poisson brackets requires the introduction of independent local coordinates on $F_1^{\pm}$. Each choice of such coordinates, however, would destroy the manifest conformal symmetry of the formalism. We shall proceed in an alternative way, by first constructing the conformal group generators J_{ab} which give rise to a complete set of observables (and whose Poisson brackets are determined by the structure constants of the conformal Lie algebra).

We start by writing down the Liouville operators $L_{J_{ab}}$ which are equal to minus the infinitesimal operators of the representation of SU(2,2), acting on functions of ζ and $\tilde{\zeta}$.

We have

$$L_{J_{ab}} = \frac{\partial}{\partial\zeta}\gamma_{ab}\zeta - \tilde{\zeta}\gamma_{ab}\frac{\partial}{\partial\tilde{\zeta}} \tag{3.3}$$

where γ_{ab} are the 4×4 matrix-generators of SU(2,2) [given by (1.39)].

It is not difficult to verify that the operators (3.3) form an antirepresentation of the conformal Lie algebra under commutation. We shall use the following basic property of Liouville operators (see Sect.1.2 of the lecture notes [T3]):

$$\omega_s(L_f,L_g) = L_g f = -L_f g(=\{f,g\}) \quad . \tag{3.4}$$

For $f = J_{ab}$, $g = J_{cd}$ (3.1) gives

$$\omega_s(L_{J_{ab}},L_{J_{cd}}) = \frac{2is}{\tilde{\zeta}\zeta}\,\zeta(\gamma_{ab}\Pi_\zeta\gamma_{cd} - \gamma_{cd}\Pi_\zeta\gamma_{ab})\zeta$$

$$= \frac{2is}{\tilde{\zeta}\zeta}\,\tilde{\zeta}[\gamma_{ab},\gamma_{cd}]\zeta \quad .$$

Combining this result with (3.4) we find

$$J_{ab} = \frac{2is}{\tilde{\zeta}\zeta}\,\tilde{\zeta}\gamma_{ab}\zeta \quad , \tag{3.5}$$

$$\{J_{ab},J_{cd}\} = \frac{2is}{\tilde{\zeta}\zeta}\,\tilde{\zeta}[\gamma_{ab},\gamma_{cd}]\zeta \quad . \tag{3.6}$$

The knowledge of the basic observables J_{ab} provides a complete characterization of the mechanical system under consideration. We shall demonstrate that the conformal generators (3.5) describe a zero mass particle of helicity s. The simplest way to verify the defining relations for such a system,

$$p^2 = 0 \quad , \tag{3.7}$$

$$W^\lambda \equiv \frac{1}{2}\,\varepsilon^{\lambda\mu\nu\rho}J_{\mu\nu}P_\rho = sP^\lambda(\varepsilon^{0123} = -\varepsilon_{0123} = 1) \quad , \tag{3.8}$$

is to express J_{ab} (in particular, P_μ and $J_{\mu\nu}$) in terms of the 2-component Lorentz spinors π and χ introduced by (2.24) [in the Cartan basis (1.37)]. We find

$$P_\mu = J_{\mu 6} - J_{\mu 5} = \frac{2s}{\tilde{\zeta}\zeta}\,\bar{\pi}\sigma_\mu\pi \quad \text{with} \quad \tilde{\zeta}\zeta = \chi\pi + \bar{\pi}\bar{\chi} \tag{3.9a}$$

$$K_\mu = J_{\mu 6} + J_{\mu 5} = \frac{2s}{\tilde{\zeta}\zeta}\,\chi\sigma_\mu\bar{\chi} \tag{3.9b}$$

$$J_{\mu\nu} = \frac{s}{\tilde{\zeta}\zeta}\,(\chi\sigma_{\mu\nu}\pi + \bar{\pi}\sigma^*_{\mu\nu}\bar{\chi}) \quad , \qquad J_{56} = \frac{is}{\tilde{\zeta}\zeta}\,(\chi\pi - \bar{\pi}\bar{\chi}) \quad . \tag{3.10}$$

The completeness relations for the Pauli matrices

$$\tilde{\sigma}^\mu{}_{\dot{A}A}\tilde{\sigma}_{\mu\dot{B}B} = 2\varepsilon^*_{\dot{A}\dot{B}}\varepsilon_{AB} \tag{3.11a}$$

$$\sigma_\mu{}^{A\dot{A}}\sigma^{\mu B\dot{B}} = 2\varepsilon^{AB}\varepsilon^{*\dot{A}\dot{B}} \quad , \tag{3.11b}$$

where

$$(\varepsilon_{AB}) = (\varepsilon^{*\dot{A}\dot{B}}) = \begin{pmatrix} 0 & 1 \\ -1 & 0 \end{pmatrix} = (-\varepsilon^{AB}) = (-\varepsilon^*_{\dot{A}\dot{B}}) \quad , \tag{3.11c}$$

imply that the vectors (3.9) are isotropic $p^2 = 0 = K^2$, thus verifying (3.7).

The condition that the particle energy $p^0 = -p_0$ is positive provides a relation between the sign of the helicity and the type of conformal orbit in F_1; we have

$$\frac{2s}{\tilde{\zeta}\zeta} < 0 \quad (\text{for} \quad p^0 > 0) \quad . \tag{3.12}$$

The verification of (3.8) is also straightforward. For instance the zeroth component of the Pauli-Lubanski vector is given by

$$W^0 = p_1 J_{23} + p_2 J_{31} + p_3 J_{12} = - \frac{2s^2}{(\tilde{\zeta}\zeta)^2} \bar{\pi}\underline{\sigma}\pi(\chi\underline{\sigma}\pi + \bar{\pi}\underline{\sigma}\bar{\chi}) = - \frac{2s^2}{\tilde{\zeta}\zeta} \pi\bar{\pi} = sp^0 \quad . \tag{3.13}$$

Regarded as a conformal group orbit the phase space

$$\Gamma_{0s} = F_1^{-\text{signs}} \tag{3.14}$$

is highly degenerate: it is a 6-dimensional subspace of the 15-dimensional space spanned by the generators (3.5) of SU(2,2). It is, therefore, not surprising that conditions (3.7) and (3.8) are not the only quadratic identities among these generators.

Exercise 3.1. Prove the following relations among the generators (3.9) and (3.10) of SU(2,2):

$$p^\mu J_{\mu\nu} + p_\nu J_{56} = 0 \tag{3.15a}$$

$$-pK\left[=2\left(\frac{2s}{\tilde{\zeta}\zeta}\right)^2 |\chi\pi|^2\right] = 2(s^2 + J_{56}^2) \quad . \tag{3.15b}$$

[*Hint*: in verifying (3.15b) use the identity

$$-\sigma_\mu^{A\dot{A}} \tilde{\sigma}^\mu_{\dot{B}B} = 2\delta_B^A \delta_{\dot{B}}^{\dot{A}} \quad ; \tag{3.16}$$

compare with (3.11).]

3.2 Canonical Symplectic Structure on Twistor Space; a Unified Phase Space Picture for Free 0-Mass Particles

The canonical conformal invariant (contact) 1-form

$$\theta_T = \frac{i}{2} (\tilde{\zeta} d\zeta - \zeta d\tilde{\zeta}) \tag{3.17}$$

on T is a natural extension of the family of forms θ_s (3.2) [defined for $s \neq 0$ on the open subset Γ_{0s} (3.14) of $F_1 = PT$]. Indeed, we can identify Γ_{0s} with the "space of circles"

$$\Gamma_{0s} = M_{0s}/U(1) \tag{3.18}$$

where M_{0s} is the 7-dimensional "0-mass helicity s shell"

$$M_{0s} = \{\zeta \in T \ ; \ \ \tilde{\zeta}\zeta + 2s = 0\} \tag{3.19}$$

and U(1) is the group of phase transformations

$$U(1) \ni U_\alpha : \zeta \to \exp\left(i\,\frac{\alpha}{2}\right)\zeta\left[\tilde{\zeta} \to \exp\left(-i\,\frac{\alpha}{2}\right)\tilde{\zeta}\right] \quad 0 \leq \alpha \leq 4\pi \quad . \tag{3.20}$$

[Note that the equation $\tilde{\zeta}\zeta + 2s = 0$ which defines M_{0s} is consistent with the energy positivity condition (3.12).]

Clearly θ_S and θ_T coincide on M_{0s}.

The 1-form θ_T gives rise to the symplectic form

$$\omega_T = d\theta_T = i d\tilde{\zeta} \wedge d\zeta \tag{3.21}$$

on twistor space, which leads to the canonical Poisson bracket relation

$$\{\zeta^\alpha, \tilde{\zeta}_\beta\} = i\delta^\alpha_\beta \qquad (\{\zeta^\gamma, \zeta^\beta\} = 0) \quad . \tag{3.22a}$$

or, in the 2-component language,

$$\{\pi^A, \chi_B\} = i\delta^A_B \ , \ \{\bar{\chi}_{\dot{B}}, \bar{\pi}^{\dot{A}}\} = i\delta^{\dot{A}}_{\dot{B}} \quad (\{\pi^A, \bar{\pi}^{\dot{A}}\} = 0 = \{\chi_B, \bar{\chi}_B\}) \quad . \tag{3.22b}$$

The conformal group generators appear in this picture as bilinear functions of ζ and $\tilde{\zeta}$.

Exercise 3.2. Verify that the dynamical variables

$$J_{ab} = -i\tilde{\zeta}\gamma_{ab}\zeta \tag{3.23}$$

[which coincide with the generators (3.5) for $\zeta \in M_{0s}$] satisfy the Poisson bracket relations of the conformal Lie algebra:

$$\{J_{ab}, J_{cd}\} = -i\tilde{\zeta}[\gamma_{ab}, \gamma_{cd}]\zeta = \eta_{ab}J_{bd} - \eta_{ad}J_{bc} + \eta_{bd}J_{ac} - \eta_{bc}J_{ad} \tag{3.24}$$

[provided that the Poisson brackets among the ζ's are given by (3.22)].

In particular, the expressions (3.9) for p and K are substituted by the simpler formulae

$$p_\mu = -\bar{\pi}\tilde{\sigma}_\mu\pi \tag{3.25a}$$

$$K_\mu = -\chi\sigma_\mu\bar{\chi} \tag{3.25b}$$

which again lead to the relation $p^2 = 0 = K^2$. Thus T appears as the phase space of 0-mass particles of arbitrary helicity given by the variable

$$S = -\frac{1}{2}\tilde{\zeta}\zeta = -\frac{1}{2}(\chi\pi + \bar{\pi}\bar{\chi}) \quad . \tag{3.26}$$

Exercise 3.3. Verify the relation

$$W_\lambda = Sp_\lambda \tag{3.27}$$

[satisfied by the conformal generators (3.23,25)]. Show that the Liouville operator

$$L_S = \frac{i}{2}\left(\tilde{\zeta}\,\frac{\partial}{\partial\tilde{\zeta}} - \zeta\,\frac{\partial}{\partial\zeta}\right) \tag{3.28}$$

generates the group U(1) of phase transformations (3.20).

Fixing the value of the helicity, i.e. setting

$$S = s \tag{3.29}$$

where s is a given real number, amounts to restricting the variables ζ to the hypersurface M_{0s} (3.19). The restriction $\omega_{M_{0s}}$ is degenerate – it has a 1-dimensional kernel spanned by the vector fields λL_S. Hence, it gives rise to a symplectic form $\omega_{\Gamma_{0s}}$ on the factor space (3.18). It is not difficult to verify that for $s \neq 0$ the mapping

$$F_1^{-\text{signs}} \ni [\zeta] \to \left\{\exp\left(i\,\frac{\alpha}{2}\right)\zeta\right\} \in \Gamma_{0s} \qquad (\tilde{\zeta}\zeta = -2s) \tag{3.30}$$

is an isomorphism of symplectic manifolds, so that we can identify $\omega_{\Gamma_{0s}}$ with ω_{0s} (3.1). However, the present construction also applies to the case of zero helicity (which does not correspond to a conformal orbit in F_1). Thus we achieved a unified description of the free 0-mass particle phase space regarding (3.29) (for arbitrary s) as a constraint on the symplectic manifold T.

Note that the conformal generators (3.23) (considered as functions of M_{0s}) are U(1)-invariant and can consequently be regarded as functions of the factor space Γ_{0s}. Functions $f(\tilde{\zeta},\zeta)$ with this property are called observables (on M_{0s}).

Equivalently, f is an observable if it has a zero Poisson bracket with the constraint (3.29) (on M_{0s}):

$$\{f, S-s\}\big|_{M_{0s}} = 0 \quad . \tag{3.31}$$

The algebra of observables on M_{0s} is generated by the J_{ab}. There are six functionally independent observables in the neighbourhood of each point of M_{0s} which can be taken as local coordinates on Γ_{0s}.

Remark. The above phase space picture of classical zero mass particles does not include the notion of space-time coordinates. This is a reflection in the classical framework of the well-known problem of quantum mechanical localizability of the photon (concerning the conventional wisdom on this problem, see [W6]; an alternative view is developed in [A4][B1]). However, we can still explore the space time dependence of the twistor components π and χ (and of the conformal generators), since we know the action U(a) of the translation group. Indeed, for any dynamical variable $f = f(\zeta,\bar{\zeta})$ we have

$$U(a)f = \exp(-L_{ap})f = f + \{ap,f\} + \frac{1}{2}\left\{ap,\{ap,f\}\right\} + \ldots \tag{3.32}$$

where the Liouville operator L_{ap} is given [according to (3.3)] by

$$L_{ap} = i\left(\frac{\partial ap}{\partial \bar\pi}\frac{\partial}{\partial \bar\chi} - \frac{\partial ap}{\partial \pi}\frac{\partial}{\partial \chi}\right) = i\left(\bar\pi\tilde a\frac{\partial}{\partial \chi} - \frac{\partial}{\partial \bar\chi}\tilde a\pi\right) \quad . \tag{3.33}$$

We find, in particular,

$$U(a)\pi = \pi,\ U(a)\bar\pi = \bar\pi\ ;\ U(a)\chi = \chi - i\bar\pi\tilde a\ ,\ U(a)\bar\chi = \bar\chi + i\tilde a\pi \tag{3.34}$$

and hence

$$U(a)p = p\ ,\ U(a)J_{\mu\nu} = J_{\mu\nu} - a_\mu p_\nu + a_\nu p_\mu\ ,$$

$$U(a)J_{56} = J_{56} + ap\ ,\ U(a)K_\mu = K_\mu + i(\bar\pi\tilde a\sigma_\mu\bar\chi - \chi\sigma_\mu\tilde a\pi) + \bar\pi\tilde a_\mu\sigma\tilde a\pi \quad .$$

Thus χ can be regarded as a spinor field which depends linearly on x:

$$[U(a)\chi](x) = \chi(x - a) = \chi(x) - i\bar\pi\tilde a \quad , \quad \text{so that} \quad \chi(x) = \chi(0) - i\bar\pi\tilde x \quad . \tag{3.35}$$

Exercise 3.4. Prove that the function $\chi(x)$ (3.35) satisfies the *Penrose twistor equation*

$$\tilde\partial_{\dot A(B}\chi_{C)}(x) \equiv \tilde\partial_{\dot A B}\chi_C(x) + \tilde\partial_{\dot A C}\chi_B(x) = 0 \quad , \tag{3.36}$$

(where $\tilde\partial = [(\partial/\partial x_\mu)\tilde\sigma^\mu]$).

3.3 The Phase Space of Spinless Positive Mass "Conformal Particles"

We shall now study the conformal invariant symplectic structure on the 8-dimensional flag manifolds $F_2^{\pm\pm} \simeq T_\pm$ of Sect.2.3. The symplectic form is fixed by $SO_0(4,2)$ ($\simeq SU(2,2)/Z_2$) invariance (and energy positivity) up to a (positive) factor. In terms of the homogeneous coordinates [Z] [on the projective quadric $P(Q_C)$] (2.7-8) it can be written as

$$\omega_0 = -i\rho\bar d\wedge d\ \ln|\bar ZZ| = \frac{-i\rho}{\bar ZZ}d\bar Z\wedge\Pi_Z dZ = \frac{-i\rho}{\bar ZZ}\left(d\bar Z\wedge dZ - \frac{Zd\bar Z\wedge\bar ZdZ}{\bar ZZ}\right) \tag{3.37}$$

[d and $\bar d$ standing for differentiation with respect to Z and $\bar Z$, respectively; $\Pi_Z = 1 - (Z\otimes\bar Z/\bar ZZ)$].

Going to the independent coordinates $z^\mu(= x^\mu + iy^\mu) = \frac{1}{\kappa}Z^\mu$ we find

$$\omega_0 = -i\rho d\bar z^\mu\wedge dz\ \frac{\nu\partial^2}{\partial\bar z^\mu\partial z^\nu}\ln(z - \bar z)^2 = \rho\frac{r_{\mu\nu}(y)}{y^2}dx^\mu\wedge dy^\nu \quad , \tag{3.38a}$$

where

$$r_{\mu\nu}(y) = \eta_{\mu\nu} - 2\frac{y_\mu y_\nu}{y^2} \qquad (\text{so that}\quad r^\mu_\lambda r^\lambda_\nu = \delta^\mu_\nu) \quad . \tag{3.38b}$$

According to the *Darboux theorem* for every symplectic form ω there exist *canonical* local *coordinates* (q^α, p_β) such that $\omega = dq^\alpha\wedge dp_\alpha$. For the form ω_0 (3.33) a natural choice of canonical coordinates is

$$q^\mu = x^\mu \ , \quad p_\nu = \rho \frac{y}{y^2} \quad \left(y_\nu = \frac{\rho p_\nu}{p^2} \right) \ . \tag{3.39}$$

$$\omega_0 = dx^\mu \wedge dp_\mu \tag{3.40}$$

is a consequence of the identity

$$d \frac{y_\mu}{y^2} = \frac{r_{\mu\nu}(y)}{y^2} dy^\nu \quad .$$

Our next task is to find the so-called "momentum map" (in the terminology of [L3] [O1,2]), i.e. the expression for the SU(2,2) generators J_{ab} in terms of the z-variables. Every antisymmetric tensor J_{ab} on PQ is proportional to $i(Z_a\bar{Z}_b - Z_b\bar{Z}_a)/\bar{Z}Z$. The coefficient is determined by the requirement that J_{ab} provide a Poisson bracket realization of the conformal Lie algebra.

The result is

$$J_{ab} = \frac{-i}{\bar{Z}Z}(\bar{Z}_a Z_b - \bar{Z}_b Z_a) \quad . \tag{3.41}$$

Let us do again the exercise showing that

$$\{J_{ab}, J_{cd}\} = \eta_{ab}J_{bd} - \eta_{ad}J_{bc} + \eta_{bd}J_{ac} - \eta_{bc}J_{ad} \tag{3.42}$$

[cf (1.31)].

Exercise 3.5. Use (3.32) for $f = J_{ab}$ and $X = \lambda Z\wedge\partial + \mu\bar{Z}\wedge\bar{\partial}$ to show that

$$L_{J_{ab}} = Z_a\partial_b - Z_b\partial_a + \bar{Z}_a\partial_b - \bar{Z}_b\bar{\partial}_a \quad \left(\partial_a = \frac{\partial}{\partial Z^a} \ , \quad \bar{\partial}_a = \frac{\partial}{\partial \bar{Z}^a} \right) \ . \tag{3.43}$$

Equations (3.33,41, and 43) now imply (3.42).

Another (more conventional) way to verify (3.41) [and (3.42)] is to express J_{ab} in terms of the independent canonical coordinates x and p. We have

$$J_{\mu\nu} = \frac{-2i\rho}{(z-\bar{z})^2}(z_\mu\bar{z}_\nu - z_\nu\bar{z}_\mu) = x_\mu p_\nu - x_\nu p_\mu \ , \quad J_{56} = -xp \ , \tag{3.44a}$$

$$J_{\mu 6} - J_{\mu 5} = \frac{-2i\rho}{(z-\bar{z})^2}(\bar{z}_\mu - z_\mu) = p_\mu \ , \quad K_\mu = J_{\mu 6} + J_{\mu 5}$$
$$= x^2 p_\mu - 2(xp)x_\mu - \frac{\rho^2}{p^2} p_\mu \quad . \tag{3.44b}$$

The SO(4,2) Lie algebra relations now appear as a straightforward consequence of the canonical commutation relations among x^μ and p_ν.

We see, in particular, that p_μ (3.39) plays the role of particle momentum. Since $y^2 < 0$ in both T_+ and T_-, energy positivity implies $\rho y^0 < 0$. We shall choose in what follows $\rho > 0$, identifying the classical phase space of a positive energy spinless "conformal particle" with the backward tube T_- (F_2^{--}). [Note that T_- is the analyticity domain for quantum mechanical particle wave functions satisfying the spectrum condition — i.e., having momentum space support in the forward cone $p^0 \geqslant |\underline{p}|$. As we shall see in Sect.3.4 below, the choice $\rho > 0$ is also natural in the study of positive energy (and positive mass) spinning particles.]

The term "conformal particle" indicates that the (positive) value of the mass operator $-p^2$ is not fixed. A constraint of the type $p^2 + m^2 = 0$ is only conformal invariant for $m = 0$. [We leave it as an exercise to the reader to verify that the zero-mass shell $p^2 = 0 (p \neq 0)$ intersects the boundary of T_- at infinity on a 5(real) dimensional cone which is given (in homogeneous coordinates) by

$$(Z - \bar{Z})^2 = -2\bar{Z}Z = 0, Z^5 + Z^6 = 0 \ .]$$

The description of a free massive particle (with fixed mass $m > 0$) requires breaking the conformal symmetry to its Poincaré subgroup. The larger "conformal phase space" described here can serve as a playground in the constraint Hamiltonian description of the massive particles (see [T3]).

3.4 The 10-Dimensional Phase Space of a Timelike Spinning Particle

As we have seen in Sect.2.4 the flag $F_{1,2}$ can be parametrized by the set of pairs $([Z],[\zeta]; Z \in Q_{\mathbb{C}}, \zeta \in T)$ satisfying

$$Z\check{\beta}\zeta = 0 = \tilde{\zeta}\beta Z \qquad (\tilde{\zeta} = \bar{\zeta}A, Z_a Z^a = 0) \ . \tag{3.45}$$

The (open) SU(2,2)-orbits $F^{\pm}_{1,2} = \{([Z],[\zeta]) \in F_{1,2}; Z \in T_{\pm}\}$ can be parametrized by finite points $(Z^\mu = \kappa z^\mu, \ \kappa \neq 0)$ of the complex Minkowski space CM. For such points (3.45) and its general solution may be written in the form [we use the Lorentz-Cartan basis (1.37) and (2.25)]

$$\begin{pmatrix} -z^2 & -i\underset{\sim}{z} \\ -i\tilde{z} & -1 \end{pmatrix} \zeta = 0 \Rightarrow \zeta = \begin{pmatrix} \pi \\ -i\tilde{z}\pi \end{pmatrix} \quad ; \quad \tilde{\zeta} = (i\bar{\pi}\bar{\tilde{z}}, \bar{\pi}) \ , \quad \pi = (\pi^A, A = 1,2) \ . \tag{3.46}$$

Thus $F^{\pm}_{1,2}$ can be identified with the complex 5-dimensional manifold $T_{\pm}(\ni z) \times \mathbb{C}P_1(\ni[\pi])$.

The most general (real) conformal invariant symplectic form on $F^{\pm}_{1,2}$ can be presented as the sum of the forms (3.37) and (3.25):

$$\omega_s^+ = \omega_0 + \omega_s = d\left(\rho \, \frac{\bar{Z}dZ - Zd\bar{Z}}{2i\bar{Z}Z} + s \, \frac{\tilde{\zeta}d\zeta - \zeta d\tilde{\zeta}}{i\tilde{\zeta}\zeta}\right) \ . \tag{3.47}$$

The superscript + on ω indicates that we are going to arrange the signs of ρ and $y^0 = \text{Im}\, z^0$ (for given s) in such a way that the energy $J_{05} - J_{06} (= -p_0)$ be positive. If we also wish to identify the (real) parameter s with the spin of the particle (and hence set $s > 0$) then we are led to study the orbit $F^{\pm}_{1,2}$ (for which $y^0 < 0$) with $\rho > 0$.

Following the approach of Sect.3.2 we shall extend the symplectic structure (3.47) to a (real) 12-dimensional manifold Γ_+ replacing the compact manifold $\mathbb{C}P_1$ of projective spinors $[\pi]$ by $\mathbb{C}^2$; the physical phase space

$$\Gamma_s^+ = (F_{1,2}^-, \omega_s^+, s > 0) \tag{3.48}$$

will then be identified with a factor space of the 11-dimensional spin shell

$$M_s^+ = \{(z,\tilde{\pi}) \in T\_x\mathbb{C}^2 \ , \ \tilde{\zeta}\zeta = 2\bar{\pi}\tilde{y}\pi = -2s\} \ . \tag{3.49}$$

For ζ satisfying (3.46 and 49) the second term in the right-hand side of (3.47) goes into

$$(\omega_s \to) \ d\,\frac{i}{2}\,(\tilde{\zeta}d\zeta - \zeta d\tilde{\zeta}) = d[\bar{\pi}d\tilde{x}\pi + i(\bar{\pi}\tilde{y}d\pi - (d\bar{\pi})\tilde{y}\bar{\pi})] \quad .$$

Inserting this expression in (3.47) and using

$$\omega_0 = d\left(\frac{\rho y}{-y^2}\,dx\right) \quad [\text{see (3.38)}]$$

we find the following extension ω^+ of the form ω_s^+ to Γ_+:

$$\omega^+ = d[-\,pdx + i(\bar{\pi}\tilde{y}d\pi - (d\bar{\pi})\tilde{y}\pi)] \tag{3.50}$$

where

$$p = \frac{\rho y}{y^2} - k \ , \quad k_\mu \equiv \bar{\pi}\sigma_\mu\pi \quad (k^2 = 0 \ , \ k_0 > 0) \quad . \tag{3.51}$$

We can express, conversely, y as a function of p and k

$$y = \rho\,\frac{p+k}{(p+k)^2} \quad . \tag{3.52}$$

Exercise 3.6. For k given by (3.51) establish the relations

$$\underset{\sim}{k}^{A\dot{B}} = -2\pi^A\bar{\pi}^{\dot{B}} \ , \tag{3.53}$$

$$\tilde{k}\pi = 0 = \bar{\pi}\tilde{k} \quad . \tag{3.54}$$

[*Hint*: in proving (3.53) use the summation formula

$$\sigma_\mu^{A\dot{A}}\sigma^\mu_{B\dot{B}} = -2\delta^A_B\delta^{\dot{A}}_{\dot{B}} \quad .]$$

We can rewrite the form ω^+ (3.50) in terms of the independent variables x, p, π and $\bar{\pi}$ as

$$\omega^+ = d\left[-\,p\,dx + i\rho\,\frac{\bar{\pi}\tilde{p}d\pi - (d\bar{\pi})\tilde{p}\pi}{(p+k)^2}\right] \tag{3.55a}$$

$$= dx\wedge dp + \frac{i\rho}{(p+k)^4}[2p^2 d\bar{\pi}\wedge(\tilde{p} + \tilde{k})d\pi + \bar{\pi}\tilde{p}dp\wedge(\tilde{p} + \tilde{k})d\pi + d\bar{\pi}\wedge(\tilde{p} + \tilde{k})dp\tilde{p}\pi] \tag{3.55b}$$

$$= \frac{1}{2}\,(dx,dp,d\bar{\pi},d\pi)\begin{pmatrix} 0 & 1 & 0 & 0 \\ -1 & 0 & -i\rho\,\frac{(\tilde{p}+\tilde{k})\sigma\tilde{p}\pi}{(p+k)^4} & i\rho\,\frac{\bar{\pi}\tilde{p}\sigma(\tilde{p}+\tilde{k})}{(p+k)^4} \\ 0 & i\rho\,\frac{(\tilde{p}+\tilde{k})\sigma\tilde{p}\pi}{(p+k)^4} & 0 & 2i\rho p^2\,\frac{\tilde{p}+\tilde{k}}{(p+k)^4} \\ 0 & -i\rho\,\frac{\bar{\pi}\tilde{p}\sigma(\tilde{p}+\tilde{k})}{(p+k)^4} & -2i\rho p^2\,\frac{\tilde{p}+\tilde{k}}{(p+k)^4} & 0 \end{pmatrix}\begin{pmatrix} dx \\ dp \\ d\bar{\pi} \\ d\pi \end{pmatrix} . \tag{3.55c}$$

The inverse matrix, call it Ω, to the matrix in (3.55c) provides the expressions for the basic Poisson brackets in Γ_+:

$$\{(x,p,\bar{\pi},\pi),\begin{pmatrix} x \\ p \\ \bar{\pi} \\ \pi \end{pmatrix}\} = {}^t\Omega \tag{3.56}$$

(the superscript to the left of Ω standing, as usual, for transposition).

Exercise 3.7. Show that

$$\Omega = \begin{pmatrix} \frac{s_{\mu\nu}}{p^2} & -1 & \frac{\bar{\pi}\tilde{p}\sigma_\mu}{2p^2} & \frac{\sigma_\mu\tilde{p}\pi}{2p^2} \\ 1 & 0 & 0 & 0 \\ \frac{\bar{\pi}\tilde{p}\sigma_\nu}{-2p^2} & 0 & 0 & -\frac{i(p+k)^2}{2\rho p^2}\,(\underset{\sim}{p}+\underset{\sim}{k}) \\ \frac{\sigma_\nu\tilde{p}\pi}{-2p^2} & 0 & \frac{i(p+k)^2}{2\rho p^2}\,(\underset{\sim}{p}+\underset{\sim}{k}) & 0 \end{pmatrix} \tag{3.57}$$

where

$$s_{\mu\nu} = -\frac{\rho}{2(p+k)^2}\,\bar{\pi}(\tilde{p}\sigma_{\mu\nu} + \sigma^*_{\mu\nu}\tilde{p})\pi = \frac{i}{2(p+k)^2}\,\bar{\pi}(\tilde{\sigma}_\mu\tilde{p}\sigma_\nu - \tilde{\sigma}_\nu p\tilde{\sigma}_\mu)\pi \tag{3.58a}$$

$$\sigma_{\mu\nu} = \frac{i}{2}\,(\sigma_\mu\tilde{\sigma}_\nu - \sigma_\nu\tilde{\sigma}_\mu)\ , \quad \sigma^*_{\mu\nu} = \frac{i}{2}\,(\tilde{\sigma}_\mu\sigma_\nu - \tilde{\sigma}_\nu\sigma_\mu)\quad . \tag{3.58b}$$

From (3.56 and 57) we find the following Poisson brackets for the basic variables in Γ_+:

$$\{x^\mu,x^\nu\} = \frac{s^{\mu\nu}}{-p^2}\ ,\ \{x^\mu,p_\nu\} = \delta^\mu_\nu\ ,\ \{x^\mu,\bar{\pi}\} = \frac{\bar{\pi}\tilde{p}\sigma^\mu}{-2p^2}\ ,\ \{x^\mu,\pi\} = \frac{\sigma^\mu\tilde{p}\pi}{-2p^2} \tag{3.59a}$$

$$\{p_\mu,p_\nu\} = 0 = \{p_\mu,\pi\} = \{p_\mu,\bar{\pi}\} \tag{3.59b}$$

$$\{\pi^{-\dot{A}},\pi^{A}\} = i\,\frac{(p+k)^2}{2\rho p^2}\,(\underset{\sim}{p}+\underset{\sim}{k})^{A\dot{A}} \quad (\{\pi^{A},\pi^{B}\} = 0 = \{\pi^{-\dot{A}},\pi^{-\dot{B}}\}) \quad . \tag{3.59c}$$

The skew symmetric tensor $s_{\mu\nu}$ (3.58) can be identified with the spin part of angular momentum (as suggested by our notation). That is seen if we take for the conformal generators the sum of the expressions (3.27 and 41):

$$J_{ab} = -\frac{i\rho}{\bar{Z}Z}\,(\bar{Z}_a Z_b - \bar{Z}_b Z_a) - i\tilde{\zeta}\gamma_{ab}\zeta \quad . \tag{3.60}$$

We find, in particular, the following expressions for the generators of the Poincaré subalgebra

$$J_{\mu 6} - J_{\mu 5} = p_\mu \tag{3.61a}$$

$$J_{\mu\nu} = x_\mu p_\nu - x_\nu p_\mu + s_{\mu\nu} \tag{3.61b}$$

[where $s_{\mu\nu}$ is given by (3.58)].

In order to verify the standard Poisson bracket relations of the Poincaré Lie algebra, we shall evaluate [using (3.58,59)] the various Poisson brackets of $s_{\mu\nu}$. Albeit straightforward, this calculation is rather lengthy. It can be simplified if we make use of the identities

$$s_{\mu\nu}p^\nu = 0 = s_{\mu\nu}k^\nu \quad . \tag{3.62}$$

They suggest that the bivector $s_{\mu\nu}$ is proportional to $\varepsilon_{\mu\nu\kappa\lambda}p^\kappa k^\lambda$. Indeed, introducing the lightlike vector

$$\ell = \frac{\rho k}{-(p+k)^2} \quad (\ell^2 = 0\ ,\ \ell_0 > 0) \tag{3.63}$$

(which has the dimension of x), we can write

$$s_{\kappa\lambda} = \varepsilon_{\kappa\lambda\mu\nu}p^\mu\ell^\nu \ , \qquad {}^*s_{\mu\nu} = \frac{1}{2}\,\varepsilon_{\mu\nu\kappa\lambda}s^{\kappa\lambda} = (\ell\wedge p)_{\mu\nu} \tag{3.64a}$$

$$W_\mu = {}^*s_{\mu\nu}p^\nu = p^2\ell_\mu - (p\ell)p_\mu \quad . \tag{3.64b}$$

Exercise 3.8. Verify the Poisson bracket relations

$$\{x^\mu,k^\nu\} = \frac{k^\mu p^\nu - k^\nu p^\mu - \eta^{\mu\nu}pk}{-p^2} \tag{3.65a}$$

$$\{x^\mu,(p+k)^2\} = 2\,\frac{(p+k)^2}{p^2}\,p^\mu \quad , \tag{3.65b}$$

and use them to deduce

$$\{x^\mu,\ell^\nu\} = \frac{\ell^\mu p^\nu + p^\mu\ell^\nu - \eta^{\mu\nu}\ell p}{-p^2} = -\{\ell^\mu,x^\nu\} \quad ; \tag{3.66a}$$

show furthermore that

$$\{\ell^\mu,\ell^\nu\} = \frac{s^{\mu\nu}}{-p^2} \ . \tag{3.66b}$$

Using the above exercise, it is not difficult to derive the relations

$$\{x^\lambda,s^{\mu\nu}\} = \frac{p^\mu s^{\lambda\nu} - p^\nu s^{\lambda\mu}}{-p^2} \ , \tag{3.67a}$$

$$\{s_{\kappa\lambda},s_{\mu\nu}\} = \Pi_{\kappa\mu}s_{\lambda\nu} - \Pi_{\kappa\nu}s_{\lambda\mu} + \Pi_{\lambda\nu}s_{\kappa\mu} - \Pi_{\lambda\mu}s_{\kappa\nu} \ , \tag{3.67b}$$

$$\Pi_{\mu\nu} = \eta_{\mu\nu} - \frac{p_\mu p_\nu}{p^2} \ . \tag{3.67c}$$

[Equation (3.67b) yields a simple check in the centre of mass frame, in which $p = (\sqrt{-p^2},\underline{0})$ and $s_{0\nu} = 0 = s_{\mu 0}$.]

Exercise 3.9. Use (3.67) in order to demonstrate that $J_{\mu\nu}$ (3.61b) satisfy the Poisson bracket relations of the Lie algebra of the Lorentz group. Verify also the relations involving the Pauli-Lubanski spin-vector W:

$$\{\ell^\lambda,s_{\mu\nu}\} = \frac{W_\mu\delta^\lambda_\nu - W_\nu\delta^\lambda_\mu - p^\lambda s_{\mu\nu}}{-p^2} \ ; \tag{3.68a}$$

$$\{W_\mu,x_\nu\} = -s_{\mu\nu} = \{x_\mu,W_\nu\},\ \{W_\mu,W_\nu\} = -p^2 s_{\mu\nu} \ , \tag{3.68b}$$

$$\{W_\lambda,p_\mu\} = 0 \ ,\ \{W_\lambda,s_{\mu\nu}\} = \eta_{\lambda\mu}W_\nu - \eta_{\lambda\nu}W_\mu + s_{\mu\nu}p_\lambda \ . \tag{3.68c}$$

Having thus a hold on the conformal covariant Poisson bracket structure on Γ_+, we are ready to explore its restriction to the spin shell M_s^+ (3.49).

Exercise 3.10. Show that the relation

$$ky + s = 0 \qquad (3.49)$$

is equivalent to each of the following constraints

$$p^2\varphi_s(k,p) \equiv (\rho + 2s)kp + sp^2 = 0 \ , \tag{3.69a}$$

$$(2s + \rho)\ (p + k)^2 = \rho p^2 \ , \tag{3.69b}$$

$$p\ell = s \ ; \tag{3.69c}$$

verify that any of the constraints (3.69) implies the relation

$$W^2 + p^2 s^2 = 0 \ . \tag{3.70}$$

It is not difficult to prove that the algebra of gauge invariant observables on M_s^+ [i.e. functions on Γ_+ which have (weakly) vanishing Poisson brackets with anyone of the constraints (3.69)] is generated by the vectors $x \in M, p \in V_+ (=\{p \in M, p^0 > |\underline{p}|\})$ and ℓ, satisfying $\ell^2 = 0 = \ell p - s$; alternatively, ℓ can be substituted by W, satisfying (3.70) and $Wp = 0$.

The restriction $\omega^+|_{M_s^+}$ of the symplectic form (3.55) is degenerate. It vanishes when applied to multiples of the Hamiltonian vector field

$$L_{\varphi_s} = \frac{i\rho}{2(\rho + 2s)} \left(\pi \frac{\partial}{\partial\pi} - \bar{\pi} \frac{\partial}{\partial\bar{\pi}}\right) \quad .$$

It generates the group U(1) of phase transformations

$$\pi \rightarrow \exp[i(\alpha/2)]\pi, \quad \bar{\pi} \rightarrow \exp[-i(\alpha/2)]\bar{\pi} \quad . \tag{3.71}$$

The physical phase space Γ_s^+ is obtained from the spin shell M_s^+ by factorizing with respect to this U(1) group:

$$\Gamma_s^+ = M_s^+/U(1) \quad . \tag{3.72}$$

Remark. We see that the position variables x^μ are not canonical, since, according to (3.59a), their Poisson brackets are not identically zero. This is not an accident. As pointed out back in 1948 by *Pryce* [P16] the physical coordinates of a spinning particle do not commute (in the quantum case). This property could be derived as follows (see Chap.4 of Ref.[T3]). Consider the 14-dimensional space spanned by the Poincaré generators p_μ and $J_{\mu\nu}$ and by the 4-vector x_μ, equipped with the canonical Poisson bracket structure (($\{x^\mu,x^\nu\} = 0$, $\{x^\mu,p_\nu\} = \delta^\mu_\nu$, $\{J_{\mu\nu},p_\lambda\} = \eta_{\mu\lambda}p_\nu - \eta_{\nu\lambda}p_\mu$ etc.) The mass spin shell is characterized by two first class constraints $\varphi_1 \equiv \frac{1}{2}(m^2 + p^2) \approx 0, \varphi_2 \equiv W^2 + s^2p^2 \equiv 0$. We demand that particle world lines are independent of the choice of Lagrange multipliers λ_1 and λ_2 in the Hamiltonian constraint $H = \lambda_1\varphi_1 + \lambda_2\varphi_2 (\approx 0)$. In other words, we wish that the line $[\dot{x}]$ determined by the 4-velocity coincides with the line [p] for which the 4-momentum components play the role of homogeneous coordinates (since the two lines do coincide for $\lambda_2 = 0$). This requirement leads to some additional constraints. If we further demand that the additional constraints be Poincaré (including space reflection) invariant then we end up with the constraints $s_{\mu\nu}p^\nu = 0$ (3.62) for $s_{\mu\nu} \equiv J_{\mu\nu} - x_\mu p_\nu + x_\nu p_\mu$. These constraints are not first class and lead to a modification of the Poisson brackets of non-conserved quantities. In particular, for the Dirac brackets of the position variables we find $\{x^\mu,x^\nu\}_* = (s_{\mu\nu}/-p^2)$ in accord, with (3.59a). Using (3.66) we find that the complex variables $x^\mu + i\ell^\mu$ (just as well as the original variables z^μ) have zero Poisson brackets among themselves and behave as a 4-coordinate under Poincaré transformations but their Poisson brackets with the complex conjugate 4-vector do not vanish.

3.5 The 12-Dimensional Phase Space $F^-_{1,2,3}$

We now turn to the most general symplectic flag manifold $F^-_{1,2,3}$ (cf. the end of Sect.2.3).

A point in $F^-_{1,2,3}$ can be labelled by a triple $[\zeta_-][Z][\zeta_+]$ satisfying

$$\tilde{\zeta}_+ Z\beta = 0 = Z\overset{\vee}{\beta}\zeta_- \ , \quad [Z] \in T_- \ (=F_2^{--}) \ . \tag{3.73}$$

Since the backward tube T_- is diffeomorphic to an open set in $\mathbb{C}M$ (which does not contain points at infinity) we can parametrize it by the independent (Cartesian) coordinates $z = x + iy \in T_-$ related to the homogeneous coordinates $[Z]$ by

$$Z^\mu = \kappa z^\mu (k \neq 0) \ z = x + i\ y \ , \ -y^0 > |\mathbf{y}| \ . \tag{3.74}$$

Using the Cartan basis (1.37) we can rewrite (3.73) in the form

$$\left(\frac{1}{\kappa}\,\tilde{\zeta}_+ Z\beta = \right)\tilde{\zeta}_+\begin{pmatrix} \mathbb{1} & -i\tilde{z} \\ -i\tilde{z} & z^2 \end{pmatrix} = 0 \ , \ \left(\frac{1}{\kappa}\,Z\beta\zeta_- = \right)\begin{pmatrix} -z^2 & -i\tilde{z} \\ i\tilde{z} & -\mathbb{1} \end{pmatrix}\zeta_- = 0 \ . \tag{3.75}$$

In accord with (3.46) the general solution of (3.75) (with respect to $\zeta_\pm$) is

$$\zeta_\pm = \begin{pmatrix} \pi_\pm \\ (-i\tilde{x} \mp \tilde{y})\pi_\pm \end{pmatrix} = T(x)\begin{pmatrix} \pi_\pm \\ \mp\tilde{y}\pi_\pm \end{pmatrix} , \qquad \text{where} \qquad T(x) = \begin{pmatrix} \mathbb{1} & 0 \\ -i\tilde{x} & \mathbb{1} \end{pmatrix} \tag{3.76a}$$

or

$$\zeta_- = \begin{pmatrix} \pi_- \\ -\tilde{z}\pi_- \end{pmatrix} \ \tilde{\zeta}_+ = (\bar{\pi}_+ i\tilde{\bar{z}}, \bar{\pi}_+) \ . \tag{3.76b}$$

Proceeding in analogy with the treatment of the 10-dimensional phase space Γ_s^+ we introduce the (real) 14-dimensional manifold

$$M_{s_+s_-} = \left\{(z,\pi_+,\pi_-) \in T_- \times \mathbb{C}^2 \times \mathbb{C}^2 \ ; \quad \tilde{\zeta}_\pm\zeta_\pm = \mp 2\bar{\pi}_\pm\tilde{y}\pi_\pm = \pm 2S_\pm\right\} \quad (S_\pm > 0) \tag{3.77}$$

[that is the counterpart of M_s^+ (3.49)] equipped with the closed 2-form

$$\omega = d\left[\rho\,\frac{y}{-y^2}\,dx + \frac{i}{2}\,(\tilde{\zeta}_+ d\zeta_+ - \zeta_+ d\tilde{\zeta}_+ + \tilde{\zeta}_- d\zeta_- - \zeta_- d\tilde{\zeta}_-)\right]$$

$$= d[-\,pdx - i(\bar{\pi}_+\tilde{y}d\pi_+ - (d\bar{\pi}_+)\tilde{y}\pi_+ - \bar{\pi}_-\tilde{y}d\pi_- + (d\bar{\pi}_-)\tilde{y}\pi_-)] \tag{3.78}$$

where

$$p = \rho\,\frac{y}{y^2} - K \ , \quad K = k_+ + k_- \ , \quad k^\mu_\pm = \bar{\pi}_\pm\tilde{\sigma}^\mu\pi_\pm \quad (-k^0_\pm > 0) \tag{3.79a}$$

or

$$y = \rho\,\frac{p + K}{(p + K)^2} \ . \tag{3.79b}$$

The constraints (3.77) and the completeness relations (3.11a) imply the identities

$$k_+^2 = 0 = k_-^2 \ , \quad \tilde{k}_\pm\pi_\pm = 0 \ , \quad k_+k_- = -2|\pi_+\varepsilon\pi_-|^2 \ , \tag{3.80a}$$

$$yk_\pm = -s_\pm \ , \quad py = \rho + s_+ + s_- \ , \tag{3.80b}$$

or

$$kK = 0 = K^2 + k^2 \quad \text{for} \quad k = k_+ - k_- \ , \tag{3.81a}$$

$$(s_+ - s_-)K(p + K) = (s_+ + s_-)pk \quad , \quad \rho K(p + K) + (s_+ + s_-)(p + K)^2 = 0 \quad . \tag{3.81b}$$

The inequalities coming from $(p + K)^2 < 0$, $K^2 \leqslant 0$, $p(K \pm k) > 0$ together with (3.81) imply that the variable

$$K_\perp^2 = K^2 - \frac{(KP)^2}{P^2} \tag{3.82}$$

varies in the range

$$\frac{\rho^2(s_+ - s_-)^2}{(\rho + 2s_+)^2(\rho + 2s_-)^2} \leqslant \frac{K_\perp^2}{-P^2} \leqslant \left(\frac{s_+ + s_-}{\rho + 2s_+ + 2s_-}\right)^2 \tag{3.83}$$

$$\frac{K_\parallel}{\sqrt{-p^2}} \equiv \frac{Kp}{-p^2} = \frac{\rho + 2s_+ + 2s_-}{2(\rho + s_+ + s_-)} - \sqrt{\left(\frac{\rho + 2(s_+ + s_-)}{2(\rho + s_+ + s_-)}\right)^2 - \frac{s_+ + s_-}{\rho + s_+ + s_-} + \frac{K_\perp^2}{m^2}} \quad . \tag{3.84}$$

Exercise 3.10. Verify that for $s_+ = 0$, $s = s(\pi_+ = 0, K = -k = k_-)$. Equations (3.77-84) reproduce the characteristic equations of the preceding section for the spin shell M_s^+. In particular, (3.83 and 84) imply

$$\sqrt{\frac{K_\perp^2}{-p^2}} = \frac{Kp}{-p^2} = \frac{s}{\rho + 2s}$$

in accord with (3.69a).

Defining with intrinsic angular momentum of the system under consideration as

$$s_{\mu\nu} = J_{\mu\nu} - L_{\mu\nu} \; [= J_{\mu\nu} - (x_\mu p_\nu - x_\nu p_\mu)]$$

$$= - i\tilde{\psi}_+\gamma_{\mu\nu}\psi_+ - i\tilde{\psi}_-\gamma_{\mu\nu}\psi_- \qquad \text{for} \qquad \psi_\pm = T(-x)\zeta_\pm = \begin{pmatrix} \pi_\pm \\ -\\ +\tilde{y}\pi_\pm \end{pmatrix} \quad , \tag{3.85}$$

we obtain [in parallel with (3.64a)]

$$^*S_{\mu\nu} = k_\mu y_\nu - k_\nu y_\mu \tag{3.86}$$

so that [using (3.80b and 79)]

$$W_\mu = {}^*S_{\mu\nu}p^\nu = (\rho + s_+ + s_-)k_\mu + (s_+ - s_-)(p_\mu + K_\mu) \quad . \tag{3.87}$$

Thus, the spin-square

$$S^2 = \frac{W^2}{-p^2} = (s_+ - s_-)^2 + (\rho + 2s_+)(\rho + 2s_-)\,\frac{K^2}{p^2} \tag{3.88}$$

is not a constant on $M_{s_+s_-}$. In order to find its range we observe that (3.83,84) imply

$$0 \leqslant \frac{1}{p^2}\,K^2 \leqslant \frac{4s_+s_-}{(\rho + 2s_+)(\rho + 2s_-)} \quad . \tag{3.89}$$

Hence,

$$(s_+ - s_-)^2 \leqslant S^2 \leqslant (s_+ + s_-)^2 \quad ; \tag{3.90}$$

this corresponds (in the classical picture) to the spin content of the general positive energy unitary irreducible ray representations of the conformal group, classified by *Mack* [M4].

The 12-dimensional phase space $F^-_{1,2,3}$ can be identified with the factor space

$$\Gamma_{s_+s_-} = M_{s_+s_-}/U_+(1) \times U_-(1) \tag{3.91}$$

where $U_\pm(1)$ are the groups of phase transformations $\pi_\pm \to \exp[i(\alpha_\pm/2)]\pi_\pm$ generated by the Hamiltonian vector fields of the two constraints defining $M_{s_+s_-}$ (3.77) which span the kernel of the form ω (3.78) [in full analogy with the identification of $F_{1,2}$ with Γ^+_s (3.72)]. It describes the classical phase space of a conformal particle of a variable mass ($0 < -p^2 < \infty$) and spin [in the range (3.90)].

4. Twistor Description of Classical Zero Mass Fields

4.1 Quantization of a Zero Mass Particle System: The Ladder Representations of U(2,2)

To quantize a classical dynamical system, means to construct a (Hermitian) Hilbert space representation of a given subalgebra of the infinite dimensional Lie algebra of Poisson brackets which contains the basic observables.[1] (A more systematic point of view on quantization, that uses the notion of deformation of algebraic structures is expounded in Ref. [B1].) In the case of a 0-mass particle system, we shall start with the canonical symplectic structure of Sect.3.2 and will require that the selected subalgebra includes the canonical variables ζ and $\tilde{\zeta}$ as well as the enveloping algebra of the Lie algebra of U(2,2).

Thus, we are looking, to begin with, for a pair of dual twistor operators ζ and $\tilde{\zeta}$ in a Hilbert space H, satisfying the canonical commutation relations

$$[\tilde{\zeta}_\beta, \zeta^\alpha] = \delta^\alpha_\beta \quad . \tag{4.1}$$

Remark. The simplest idea of identifying ζ and $\tilde{\zeta}$ with Fock space creation and annihilation operators fails because, if there is a "vacuum vector" $|vac\rangle$ such that $\tilde{\zeta}_\alpha |vac\rangle = 0$, then the nonzero vectors of the type $[1/2(1-A)\zeta]^\alpha |vac\rangle$ [where A is the Hermitian matrix defined by (1.42)] will have negative scalar squares. The following trick (first used by *Kursunoglu* [K6], see also [D3]) is designed to avoid the appearance of such an indefinite metric in H.

We postulate that H contains a unique state (unit ray, proportional to the vector $|0\rangle$) such that

$$\frac{1}{2}(1 - A)\zeta|0\rangle = 0 = \tilde{\zeta}\,\frac{1}{2}(1 + A)|0\rangle \quad . \tag{4.2}$$

In a basis, in which A is diagonal:

$$A = \beta_4 = \begin{pmatrix} 1 & 0 \\ 0 & -1 \end{pmatrix} \quad , \tag{4.3}$$

1 A classical example (due to Van Hove) shows that it is impossible to represent the entire Poisson bracket algebra of (smooth) functions on phase space in terms of commutators of operators on a Hilbert space (cf. [B4]).

this amounts to setting

$$\zeta = \begin{pmatrix} a^* \\ b \end{pmatrix}, \quad \tilde{\zeta} = (a,-b^*) \quad \text{where} \quad a|0> = 0 = b|0> \tag{4.4}$$

$[a=(a_1,a_2),\ b=(b_1,b_2)]$. The space H is then defined as the Hilbert space closure of the set of vectors $P(a^*,b^*)|0>$ where P is an arbitrary polynomial. {We leave it to the reader to verify that a^*, b^* and a, b satisfy, as a consequence of (4.1), the canonical commutation relations for creation and annihilation operators and that the standard Fock space construction leads to a positive metric Hilbert space H and to a unitary representation of U(2,2); see [M3] and [A5] for more details.}

The representation of U(2,2) thus obtained is reducible. It splits according to the spectrum of the (normal ordered) helicity operator

$$\hat{s} = -:C_1: = \frac{1}{2}(b^*b - a^*a)\left(=0, \pm\frac{1}{2}, \pm 1, \ldots\right) \quad . \tag{4.5}$$

In contrast to classical helicity, which can take any real value, the spectrum of $\hat{s}$ in H is discrete: s can be only integer or half integer. The space H can be split into an (infinite) direct sum of subspaces H_s of definite helicity

$$H = \bigoplus_{s=0,\pm\frac{1}{2},\pm 1,\ldots} H_s \quad . \tag{4.6}$$

The (irreducible) representation of U(2,2), realized in (any) H_s is simply reducible with respect to the maximal compact subgroup $U(2)\times U(2)$ of U(2,2).

It is not difficult to see that H_s splits into a direct sum[2]

$$H_s = \bigoplus_{\nu=0}^{\infty} H_{\nu,\nu+2s} \qquad \text{for} \quad s \geq 0$$

$$H_s = \bigoplus_{\nu=0}^{\infty} H_{\nu-2s,\nu} \qquad \text{for} \quad s < 0 \tag{4.7}$$

of $(\nu+1)(\nu+2|s|+1)$ dimensional subspaces which carry irreducible representations of $U(2)\times U(2)$. The space H_{km} consists of (all) vectors of the form $|h_{km}> = h_{km}(a_1^*,a_2^*;b_1^*,b_2^*)|0>$, where h_{km} is a homogeneous polynomial of degree k with respect to the first pair of arguments and of degree m with respect to the second one.

Exercise 4.1. Show that *Segal*'s [S1] "conformal energy operator" J_{60} has a discrete positive spectrum on H:

$$J_{60}|h_{km}> = \frac{1}{2}(k+m+2)|h_{km}>[=(\nu+|s|+1)|h_{km}>] \tag{4.8}$$

[the last equation being satisfied for either pair of integers in the right-hand side of (4.7)].

---

2 More precisely, H_s is the *topological direct sum*, i.e. the closure of the direct sum of finite dimensional subspaces in the right-hand side of (4.7).

[*Hint*: Note that in the basis (4.3,4)

$$J_{60} = \frac{1}{2} (aa^* + b^*b).]$$

Remark. The positivity of J_{60} is a consequence of the positivity of the energy P^0 for any representation of the conformal group with that property. Indeed, if P^0 has a positive spectrum, then so does K^0 which can be obtained from P^0 by a similarity transformation:

$$K^0 = \exp(i\pi J_{60})P^0 \exp(-i\pi J_{60}) \tag{4.9}$$

[$\exp(i\pi J_{60})$ represents the Weyl transformation (1.22)]. Hence, $J_{60} = \frac{1}{2} (P^0 + K^0)$ $(= -J_{06} = -\frac{1}{2} (P_0 + K_0))$ is then also positive.

The generator $\hat{s}$, being a Casimir operator, belongs to the centre of the Lie algebra of U(2,2). The (positive) operator J_{60} belongs to the centre of the maximal compact subalgebra of $U(2) \times U(2)$. The eigenvalues of these two central elements determine completely the corresponding, irreducible representation of $SU(2) \times SU(2)$ in H. [For $s \geqslant 0$ this is the representation $(\nu/2, \nu/2 + s)$ for $s < 0$, it is $(\nu/2 - s, \nu/2)$; the two (half) integer numbers in the parentheses refer to the values of the spin of the respective subgroups.] This ladder structure of the twistor representations of U(2,2) with respect to their $U(2) \times U(2)$ content has given rise to the name "*ladder representations*" for the series under consideration. They can be integrated to global representations of the 4-fold covering SU(2,2) of the conformal group of Minkowski space and form a most degenerate series of unitary irreducible representations of this group. This latter property corresponds to the low dimensionality (6) of the corresponding classical (coadjoint) orbit. As we have seen in the preceding sections a typical SU(2,2) orbit is 12 dimensional [since SU(2,2) is a 15 parameter group of rank 3, i.e. having 3 Casimir operators whose values are fixed on a given orbit], while the degenerate (lower-dimensional) orbits may have dimensions 10, 8 and 6. The ladder representations stay irreducible when restricted to the (quantum mechanical) Poincaré group ISL(2,C), since they correspond to 0-mass, fixed helicity and positive energy [which are known to determine the corresponding unitary irreducible representations of ISL(2,C)]. These properties (suggested by the classical picture) become manifest in a π-space (i.e. momentum space) realization of the ladder representations which we shall briefly describe (for more details see [M3]).

Exercise 4.2. Find the similarity transformation between the Cartan basis (1.37) of the 4×4 Clifford units β_μ and the Dirac basis in which $\beta_4(=A)$ has the diagonal form (4.3) while β_k coincide with (1.37a).

[*Answer*:

$$\beta_\mu^{Dir} = S\beta_\mu^{Car}S^{-1}$$

where

$$S = \frac{1}{\sqrt{2}}\begin{pmatrix} 1 & 1 \\ -1 & 1 \end{pmatrix}, \quad S^{-1} = S^* = {}^tS \quad .]$$

As a result of this exercise, we find the relation between the creation and annihilation operators $a^{(*)}, b^{(*)}$ and the Lorentz spinor variables π, χ (and their adjoints):

$$\begin{pmatrix} a^* \\ b \end{pmatrix} = S\begin{pmatrix} \pi \\ \chi^* \end{pmatrix} \Rightarrow a^* = \frac{1}{\sqrt{2}}(\pi + \chi^*) \quad , \quad b = (\chi^* - \pi) \quad \text{etc.} \tag{4.10}$$

The canonical commutation relations (4.1) written in terms of π and χ assume the form

$$[\chi_B, \pi^A](=i\{\chi_B, \pi^A\}) = \delta_B^A = [\pi^{A*}, \chi_{\dot{B}}^*] \quad . \tag{4.11}$$

In the π space realization, in which the operators π and π^* act as multiplication operators by π and $\bar{\pi}$, respectively, χ and χ^* are realized as the differential operators

$$\chi_A = \frac{\partial}{\partial \pi^A} \quad , \quad \chi_{\dot{B}}^* = -\frac{\partial}{\partial \bar{\pi}^{\dot{B}}} \quad . \tag{4.12}$$

In this picture, the space H consists of the square integrable functions with respect to the Lebesgue measure on $\mathbb{C}^2$:

$$\|\psi\|^2 = \frac{1}{4\pi^4} \int |\psi(\pi, \bar{\pi})|^2 d^2\pi^1 d^2\pi^2$$

$$= \int_{-2\pi}^{2\pi} \frac{d\alpha}{4\pi} \int_{V_+} (dp)_0 |\psi(p,\alpha)|^2 \quad , \quad (dp)_0 = \frac{d^3p}{(2\pi)^3 2p^0} \quad . \tag{4.13}$$

Here V_+ is the future cone $V_+ = \{p \in \mathbb{R}^4;\ p^0 = |\mathbf{p}|\}$, and we have used the change of variables

$$\mathbf{p} = \bar{\pi}\sigma\pi \quad , \quad e^{2i\alpha} = \frac{\pi^1\pi^2}{\bar{\pi}^{\dot{1}}\bar{\pi}^{\dot{2}}} \ (-2\pi < \alpha < 2\pi) \quad . \tag{4.14}$$

With the choice (4.13) for the scalar product in H the normalized lowest-weight vector $\psi_0 (=|0\rangle)$ of the generator J_{60} has the form

$$\psi_0 = 4\pi\, e^{-\bar{\pi}\pi} = 4\pi\, e^{-p^0} \quad . \tag{4.15}$$

Exercise 4.3. Show that ψ_0 is the (unique) normalized positive solution of the equations

$$a\psi_0 = \frac{1}{\sqrt{2}}\left(\bar{\pi} + \frac{\partial}{\partial \pi}\right)\psi_0 = 0 = b\psi = -\frac{1}{\sqrt{2}}\left(\pi + \frac{\partial}{\partial \bar{\pi}}\right)\psi_0 \quad . \tag{4.16}$$

Show that in terms of p and α, states of helicity s are written as

$$\psi(p,\alpha) = e^{-is\alpha}\psi_s(p) \quad .$$

Verify that $\sqrt{2}\pi^A\psi_0$ and $\sqrt{2}\bar{\pi}^{\dot{A}}\psi_0$ are (normalized) states of helicity $s = -1/2$, and $s = 1/2$, respectively [where the quantized helicity operator $\hat{s}$ is given by (4.5)].

4.2 Local Zero Mass Fields. Second Quantization

It is a remarkable general fact that the wave function of a (quantized) relativistic particle is a positive energy solution of a classical field equation with prescribed (particle) mass and spin. A free local field is a linear superposition of a positive and a negative energy solution of the field equation. The "negative energy wave functions" span the Hilbert space of the *antiladder representations* of U(2,2), which can be obtained from the construction of the preceding section by just replacing (4.3) with

$$A = -\beta_4 \quad . \tag{4.3*}$$

and setting

$$\zeta = \begin{pmatrix} \bar{a} \\ \bar{b}^* \end{pmatrix} \quad \tilde{\zeta} = (\bar{a}^*, \bar{b}) \qquad \text{where} \qquad \bar{a}|\bar{0}\rangle = 0 = \bar{b}|\bar{0}\rangle \tag{4.4*}$$

[instead of (4.4)].

Exercise 4.4. Let $\hat{s}$ and $J_{ab} = -i\tilde{\zeta}\gamma_{ab}\zeta$, and $\hat{\bar{s}}$ and $\bar{J}_{ab}$ be the generators of the ladder (L-) and the antiladder (L^*-) series of unitary irreducible representations of U(2,2), obtained from one another by substituting (4.4) with (4.4*). Show that the generators J_c of the Cartan subalgebra,

$$\{J_c\} = \{\hat{s} = 1/2(b^*b - a^*a) \; , \; J_{60} = 1/2(aa^* + b^*b) \; , \; J_{12} = 1/2(b^*\sigma_3 b - a\sigma_3 a^*) \quad ,$$

$$J_{35} = -1/2(a\sigma_3 a^* + b^*\sigma_3 b) \quad ,$$

[that are diagonal in the basis $(m_1!m_2!n_1!n_2!)^{-\frac{1}{2}}(a_1^*)^{m_1}(a_2^*)^{m_2}(b_1^*)^{n_1}(b_2^*)^{n_2}|0\rangle$] change sign under that substitution. In other words, if K is the isometric map from $H = H_L$ onto H_{L^*}, defined by $KP(a^*,b^*)|0\rangle = P(\bar{a}^*,\bar{b}^*)|0\rangle$ for any polynomial P with real coefficients (and assumed to be linear on the reals) then $KJ_c = -\bar{J}_c K$. Deduce that the ladder representation of helicity $\hat{s} = s$ is complex conjugate to the antiladder representation of helicity $\hat{\bar{s}} = -s$.

It follows, in particular, that J_{60} (and hence the energy) have negative spectrum for the L^* series.

The knowledge of the ladder (and antiladder) representations of U(2,2) provides us with a tool for constructing manifestly Lorentz covariant local 0-mass fields.

To fix the ideas we shall speak about (second) quantized fields. In this context a field $\psi(x)$ is said to be manifestly Lorentz covariant if there is a finite dimensional representation $V(\Lambda)$ of $SL(2,\mathbb{C})$ and a unitary representation $U(a,\Lambda)$ of the Poincaré group (acting in the corresponding Fock space) such that

$$U(a,\underset{\sim}{\Lambda})\psi(x)U(a,\underset{\sim}{\Lambda})^{-1} = V(\underset{\sim}{\Lambda}^{-1})\psi(\Lambda x + a) \quad . \tag{4.17}$$

Here $\Lambda = \Lambda(\underset{\sim}{\Lambda})$ is the (4×4) Lorentz matrix related to $\underset{\sim}{\Lambda}$ by

$$\underset{\sim}{\Lambda}\sigma^{\mu}\underset{\sim}{\Lambda} = \Lambda^{\mu}_{\nu}\sigma^{\nu} \quad \text{or} \quad \Lambda^{\mu}_{\nu} = -\frac{1}{2}\,\mathrm{tr}(\underset{\sim}{\Lambda}\sigma^{\mu}\underset{\sim}{\Lambda}\,\sigma_{\nu}) \tag{4.18}$$

[and ψ is a vector in the representation space of V, so that $(V\psi)^{\alpha} = V^{\alpha}_{\beta}\psi^{\beta}$]. $\psi(x)$ is local if its components commute or anticommute for spacelike separations (depending on whether 2s is even or odd). (The same concepts could be also introduced in the classical context by replacing commutators with Poisson brackets.)

The study of an arbitrary ψ of the above type can be brought to the investigation of its irreducible components. The finite dimensional representations of the Lorentz group are labelled by a pair (j_1, j_2) of non-negative integers or half-integers. *Weinberg* [W4] has demonstrated that every irreducible representation (j_1, j_2) of $SL(2,\mathbb{C})$, for which $U(a,\underset{\sim}{\Lambda})$ in (4.17) is a unitary representation of P acting in a positive metric Hilbert space carrying helicity s, is a representation of the type $(j, j+s)$. Moreover, every such field may be written as an appropriate derivation of a field transforming according to $(-s,0)$ or $(0,s)$ for $\mp s > 0$, respectively. We proceed to the description of such simple helicity fields.

For $s > 0$ we set

$$\psi(x)^{\dot{A}_1\ldots\dot{A}_{2s}} = 2^{s}\int\left[a(p,s)e^{ipx} + a^{*}(p,-s)e^{-ipx}\right]e^{i\alpha s}\bar{\pi}^{\dot{A}_1}\ldots\bar{\pi}^{\dot{A}_{2s}}\,\frac{d\pi^{1}d\bar{\pi}^{\dot{1}}d\pi^{2}d\bar{\pi}^{\dot{2}}}{(2\pi)4} \tag{4.19}$$

where a and a^{*} are annihilation and creation operators satisfying the canonical (anti)commutation relations

$$a(p,s)a^{*}(q,s') - (-1)^{2s}a^{*}(q,s')a(p,s) = (2\pi)^{3}2|\mathbf{p}|\delta(\mathbf{p}-\mathbf{q})\delta_{ss'} \tag{4.20}$$

and the transformation properties of the preceding section, while $e^{2i\alpha} = \pi^{1}\pi^{2}/\bar{\pi}^{\dot{1}}\bar{\pi}^{\dot{2}}$.

Exercise 4.5. Prove that the field ψ transforms according to the law

$$U(a,\underset{\sim}{\Lambda})\psi(x)^{\dot{A}_1\ldots\dot{A}_{2s}}U^{-1}(a,\underset{\sim}{\Lambda}) = (\bar{\underset{\sim}{\Lambda}}^{-1})^{\dot{A}_1}_{\dot{B}_1}\cdots(\bar{\underset{\sim}{\Lambda}}^{-1})^{\dot{A}_{2s}}_{\dot{B}_{2s}}\psi(\Lambda x + 0)^{\dot{B}_1\ldots\dot{B}_{2s}} \quad .$$

The 2-point function of the field ψ and its Hermitian conjugate is given by

$$\langle \overset{*}{\psi}(x)^{A_1\ldots A_{2s}} \psi(0)^{A_1\ldots A_{2s}} \rangle_0 = \int e^{ipx} \underset{\sim}{p}^{A_1\dot{A}_1} \ldots \underset{\sim}{p}^{A_{2s}\dot{A}_{2s}} (dp)_0$$

$$(dp)_0 = (2\pi)^{-3} \frac{d^3p}{2p^0} \tag{4.21}$$

here we have used (4.20) along with the annihilation property $a(p,s)|0\rangle = 0$, as well as the relations

$$\underset{\sim}{p}^{A\dot{A}} = 2\pi^A \bar{\pi}^{\dot{A}} \quad , \quad (2\pi)^{-4} d\pi^1 d\bar{\pi}^1 d\pi^2 d\bar{\pi}^2 = \frac{d\alpha}{4\pi} (dp)_0 \quad . \tag{4.22}$$

Locality follows from Lorentz invariance combined with the observation that

$$\Delta^{A_1\ldots A_{2s}\dot{A}_1\ldots\dot{A}_{2s}}(x)^{2s} = \left\langle\!\!\left\langle \left(\overset{*}{\psi}(x)^{A_1\ldots A_{2s}} \psi(0)^{\dot{A}_1\ldots\dot{A}_{2s}} - (-1)^{2s} \psi(0)^{\dot{A}_1\ldots\dot{A}_{2s}} \overset{*}{\psi}(x)^{A_1\ldots A_{2s}} \right) \right\rangle\!\!\right\rangle_0$$

$$= \int (e^{ipx} - (-1)^{2s} e^{-ipx}) \underset{\sim}{p}^{A_1\dot{A}_1} \ldots \underset{\sim}{p}^{A_{2s}\dot{A}_{2s}} (dp)_0 \tag{4.23}$$

vanishes for $x^0 = 0$, $\mathbf{x} \neq 0$.

Exercise 4.6. Setting $\underset{\sim}{p} = p^\mu \sigma_\mu$ for each (of the 2s) $\underset{\sim}{p}$ in the right-hand side of (4.23) verify that only odd powers of $p^0 (= |\mathbf{p}|)$ contribute in the integral for $x^0 = 0$. Deduce that $\Delta(0,\mathbf{x})$ is a derivative of $\delta(\mathbf{x})$.

The field $\psi(x)$ satisfies the system of first-order equations

$$\underset{\sim}{\nabla}^{A\dot{A}} \overset{*}{\varepsilon}_{AA_1} \psi^{\dot{A}_1\ldots\dot{A}_{2s}}(x) = 0 \quad , \quad \underset{\sim}{\nabla} = \partial_\mu \sigma^\mu \quad . \tag{4.24}$$

The treatment of negative helicity fields

$$\phi^{A_1\ldots A_{2|s|}}(x) = 2^{|s|} \int [a(p,s)e^{ipx} + a^*(p,-s)e^{-ipx}] e^{-i\alpha s} \pi^{A_1} \ldots \pi^{A_{2|s|}} \frac{dx}{4\pi} (dp)_0 \tag{4.25}$$

is quite similar. The only difference is that (4.24) is replaced by

$$\tilde{\nabla}_{\dot{A}A_1} \phi^{A_1\ldots A_{2|s|}}(x) = 0 \quad . \tag{4.26}$$

Exercise 4.7. Verify the validity of (4.24 and 26) for ψ and σ given by (4.19 and 25).

4.3 The Neutrino and the Photon Fields in the Twistor Picture

We end up this chapter with two physical examples: the left-handed neutrino and the photon for which a Lagrangian formalism can also be set up.

The properties of the *neutrino field* $\phi^A(x)$ can be derived from the Lagrangian

$$L = \frac{i}{2} :\left(\phi^*(x)^{\dot{A}}\tilde{\nabla}_{\dot{A}A}\, \phi(x)^A - \partial_\mu\phi^*(x)^{\dot{A}}\tilde{\sigma}^\mu_{\dot{A}A}\phi(x)^A\right): \quad ; \tag{4.27}$$

the normal product sign : : is defined, as usual, by

$$: a^*(p,s)a(q,s') : = a^*(p,s)a(q,s'),$$

$$: a(p,s)a^*(q,s') : = (-1)^{2s}a^*(q,s')a(p,s) \quad . \tag{4.28}$$

The equation of motion $\tilde{\nabla}_{\dot{A}A}\phi^A = 0$ [which is a special case of (4.26)] can be obtained by varying the Lagrangian with respect to ϕ^*. The energy momentum tensor

$$\theta_{\mu\nu}(x) = \frac{i}{4} :(\phi^*\tilde{\sigma}_\mu\partial_\nu\phi - \partial_\nu\phi^*\tilde{\sigma}_\mu\phi + \phi^*\tilde{\sigma}_\nu\partial_\mu\phi - \partial_\mu\phi^*\sigma_\nu\phi) : - \eta_{\mu\nu}L$$

leads to the following expression for the (secondly quantized) momentum generators

$$P_\mu = \int \theta_{\mu 0}d^3x = \int p_\mu\left[a^*\left(p, \frac{1}{2}\right)a\left(p, \frac{1}{2}\right) + a^*\left(p, -\frac{1}{2}\right)a\left(p, -\frac{1}{2}\right)\right](dp)_0 \quad . \tag{4.29}$$

The positivity of the expression on the right-hand side is a consequence of the Fermi character of the neutrino field [reflected in the anticommutativity of a and a^* under the sign of the normal product, cf. (4.28)].

Exercise 4.8. Verify the equal time canonical anticommutation relation

$$\left[\phi^A(0,\mathbf{x}), \phi^{*\dot{A}}(0)\right]_+ = \delta^{A\dot{A}}\delta(\mathbf{x}) \quad . \tag{4.30}$$

The Maxwell field $F_{\mu\nu}$ can be decomposed into positive and negative helicity (right-handed and left-handed) fields

$$F_{\mu\nu} = \frac{1}{\sqrt{2}}\left(F^{(+)}_{\mu\nu} + F^{(-)}_{\mu\nu}\right) \tag{4.31}$$

where

$${}^*F^{(\pm)}_{\mu\nu} = \frac{1}{2}\,\varepsilon_{\mu\nu k\lambda}F^{(\pm)k\lambda} = \pm iF^{(\pm)}_{\mu\nu} \quad . \tag{4.32}$$

For a (real) Maxwell field $F = (\mathbf{E},\mathbf{B})$ (that is a shorthand for $F^{0k}=E^k$, $F^{jk}=\varepsilon_{jk\ell}B^\ell$) we have

$$F^{(\pm)} = \frac{1}{\sqrt{2}}\,(E \mp iB, B \pm iE) \ , \quad (E = E^*, B = B^*) \quad . \tag{4.33}$$

Exercise 4.9. Verify the relations

$$F^{(+)}_{\mu\nu}(\sigma^\mu)^{A\dot{A}}(\sigma^\mu)^{B\dot{B}} = 2\varepsilon^{AB}\psi^{\dot{A}\dot{B}} \tag{4.34a}$$

$$F^{(-)}_{\mu\nu}(\sigma\mu)^{A\dot{A}}(\sigma\nu)^{B\dot{B}} = 2\phi^{AB}\varepsilon^{\dot{A}\dot{B}} \tag{4.34b}$$

where

$$\phi^{AB} = (\tau\varepsilon^{-1}\mathbf{f})^{AB} \quad , \quad \psi^{\dot{A}\dot{B}} = (\phi^*)^{\dot{A}\dot{B}} = (\varepsilon^{*-1}\tau^*\mathbf{f}^*)^{\dot{A}\dot{B}} \quad , \quad \mathbf{f} = \frac{1}{\sqrt{2}}(\mathbf{E} + i\mathbf{B}) \quad . \tag{4.34c}$$

(We recall that for the standard choice of basis $\sigma_k^{A\dot{A}}$, τ_{kB}^{A}, and $\tau_{k\dot{B}}^{*\dot{A}}$ are the same numerical matrices.)

Exercise 4.10. Prove that the free Maxwell equations $\partial_\mu F^{\mu\nu} = 0$ are equivalent to (4.24 and 26).

It is a straightforward exercise to verify that combining (4.34) with the expression (4.21) for the 2-point function one obtains the Wightman function for the free Maxwell field:

$$\langle F_{\kappa\lambda}(x)F_{\mu\nu}(0)\rangle_0 = D_{\kappa\lambda,\mu\nu} \int e^{ipx}(dp)_0 \tag{4.35a}$$

where

$$D_{\kappa\lambda,\mu\nu} = \eta_{\kappa\nu}\partial_\lambda\partial_\mu + \eta_{\lambda\mu}\partial_\kappa\partial_\nu - \eta_{\kappa\mu}\partial_\lambda\partial_\nu - \eta_{\lambda\nu}\partial_\kappa\partial_\mu \quad . \tag{4.35b}$$

Exercise 4.11. Verify that the differential operator (4.37b) satisfies the selfduality condition

$$\frac{1}{4}\varepsilon_{\kappa\lambda\kappa'\lambda'}\varepsilon_{\mu\nu\mu'\nu'}D^{\kappa'\lambda',\mu'\nu'} = D_{\kappa\lambda,\mu\nu} \quad . \tag{4.36}$$

Deduce that

$$\begin{aligned}\langle F_{\kappa\lambda}(x)F_{\mu\nu}(0)\rangle_0 &= \langle {}^*F_{\kappa\lambda}(x){}^*F_{\mu\nu}(0)\rangle_0 \\ &= \frac{1}{2}\left\langle (F^{(+)}_{\kappa\lambda}(x)F^{(-)}_{\mu\nu}(0) + F^{(-)}_{\kappa\lambda}(x)F^{(+)}_{\mu\nu}(0)\right\rangle_0 \quad .\end{aligned} \tag{4.37}$$

Exercise 4.12. Using (4.21) for the 2-point function (with $s=1$, $\psi^* = \phi$) reproduce the expression (4.35) for the 2-point function of $F_{\mu\nu}(x)$:

$$\begin{aligned}&\langle F_{\kappa\lambda}(x)F_{\mu\nu}(0)\rangle_0 \\ &= \frac{1}{8}\sigma_{\kappa\dot{A}A}\sigma_{\lambda\dot{B}B}\sigma_{\mu\dot{C}C}\sigma_{\nu\dot{D}D}\int e^{ipx}(\underset{\sim}{p}^{A\dot{C}}\underset{\sim}{p}^{B\dot{D}}\varepsilon^{\dot{A}\dot{B}}\varepsilon^{CD} + \underset{\sim}{p}^{C\dot{A}}\underset{\sim}{p}^{D\dot{B}}\varepsilon^{AB}\varepsilon^{\dot{C}\dot{D}})(dp)_0 \\ &= \int (p_\kappa p_\mu\eta_{\lambda\nu} - p_\kappa p_\nu\eta_{\lambda\mu} - p_\lambda p_\mu\eta_{\kappa\nu} + p_\lambda p_\nu\eta_{\kappa\mu})e^{ipx}(dp)_0 \quad .\end{aligned} \tag{4.38}$$

Verify the positive definiteness of the integrand.

(*Hint*: Use

$$\sigma_\lambda\tilde{\sigma}_\mu\sigma_\nu + \sigma_\nu\tilde{\sigma}_\mu\sigma_\lambda = 2\eta_{\lambda\nu}\sigma_\mu - 2\eta_{\mu\nu}\sigma_\lambda - 2\eta_{\mu\lambda}\sigma_\nu \tag{4.39a}$$

$$\tilde{\sigma}_\mu\sigma_\nu\tilde{\sigma}_\lambda + \tilde{\sigma}_\lambda\sigma_\nu\tilde{\sigma}_\mu = 2\eta_{\lambda\mu}\tilde{\sigma}_\nu - 2\eta_{\mu\nu}\tilde{\sigma}_\lambda - 2\eta_{\lambda\nu}\tilde{\sigma}_\mu \tag{4.39b}$$

as well

$$\varepsilon\, {}^t\tilde{\sigma}_\nu\, {}^t\varepsilon = \sigma_\nu \ , \ \varepsilon\, {}^t\underset{\sim}{p}\, {}^t\varepsilon = \tilde{p} \ .$$

Similarly, we verify

$$\left\langle F^{(+)}_{\kappa\lambda}(x)F^{(-)}_{\mu\nu}(0)\right\rangle_0 = \int \Big[p_\kappa p_\mu \eta_{\lambda\nu} - p_\kappa p_\nu \eta_{\lambda\mu} - p_\lambda p_\mu \eta_{\kappa\mu} + p_\lambda p_\nu \eta_{\kappa\mu} + \frac{i}{2} p^\rho(\varepsilon_{\kappa\rho\mu\nu}p_\lambda + \varepsilon_{\rho\lambda\mu\nu}p_\kappa)\Big](dp)_0 \ . \tag{4.40}$$

The asymmetry of the last term with respect to the exchange $\kappa\lambda \leftrightarrow \mu\nu$ is only apparent, because of the idenity

$$p^\rho(p_\kappa\varepsilon_{\rho\lambda\mu\nu} + p_\lambda\varepsilon_{\kappa\rho\mu\nu} + p_\mu\varepsilon_{\kappa\lambda\rho\nu} + p_\nu\varepsilon_{\kappa\lambda\mu\rho}) = 0 \tag{4.41}$$

valid for $p^2 = 0$.

To summarize: the real free Maxwell field $F_{\mu\nu}$ is given [by (4.31,33)] in terms of a complex (left-handed) symmetric spintensor field ϕ^{AB} with 2-point function

$$\left\langle \phi^{AB}(x)\phi^{*AB}(0)\right\rangle_0 = \int e^{ipx} \underset{\sim}{p}^{A\dot{A}} \underset{\sim}{p}^{B\dot{B}} (dp)_0 \ . \tag{4.42}$$

The field ϕ^{AB} can be written in tems of a (left) potential

$$\ell^A_{\dot{A}}(x) = \int \Big[\ell^{(-)}_{\dot{A}}(p)e^{ipx} + \ell^{(+)}_{\dot{A}}(p)e^{-ipx} \exp[-i(\alpha/2)]\pi^A(dp)_0 \tag{4.43}$$

where

$$i\pi^{-\dot{A}}\ell^{(-)}_{\dot{A}}(p) = e[-i(\alpha/2)]a(p,-1) \ , \ -i\pi^{-\dot{A}}\ell^{(+)}_{\dot{A}}(p) = e[-i(\alpha/2)]a^*(p,1) \ . \tag{4.44}$$

It is readily verified that the potential $\ell^A_{\dot{A}}$ so defined satisfies the equations

$$\underset{\sim}{\nabla}^{A\dot{A}}\ell^B_{\dot{A}}(x) = \phi^{AB}(x) = \frac{1}{2}\left(\underset{\sim}{\nabla}^{A\dot{A}}\ell^B_{\dot{A}} + \underset{\sim}{\nabla}^{B\dot{A}}\ell^A_{\dot{A}}\right) \equiv \underset{\sim}{\nabla}^{(A\dot{A}}\ell^{B)}_{\dot{A}} \tag{4.45a}$$

$$\tilde{\nabla}_{\dot{A}A}\ell^A_{\dot{B}}(x) = 0 \left[= \frac{1}{2}\left(\tilde{\nabla}_{\dot{A}A}\ell^A_{\dot{B}} + \tilde{\nabla}_{\dot{B}A}\ell^A_{\dot{A}}\right) \equiv \tilde{\nabla}_{(\dot{A}A}\ell^A_{\dot{B})}\right] \ . \tag{4.45b}$$

Similarly,

$$\partial_\mu \ell^*\left({}^{\dot{B}}_{A}\sigma^{\mu A\dot{A}}\right) = \phi^{*\dot{A}\dot{B}} \quad , \quad \partial_\mu \ell^*\left({}^{\dot{A}}_{B}\sigma^{\mu}_{\dot{A}A}\right) = 0 \ . \tag{4.45c}$$

Exercise 4.13. Show that the left potential $\ell^A_{\dot{A}}$ is determined from (4.45) up to a gauge transformation of the type

$$\ell^A_{\dot{A}} \to \ell^A_{\dot{A}} + \underset{\sim}{\nabla}^A_{\dot{A}}\lambda(x) \qquad \text{where} \qquad \underset{\sim}{\nabla}^A_{\dot{A}} = \nabla^{A\dot{B}}\varepsilon_{\dot{A}\dot{B}} \ . \tag{4.46}$$

One advantage of introducing the potentials ℓ and ℓ^* is the possibility they offer to write down a local Lagrangian for the electromagnetic field. We have

$$L = -\frac{1}{4} F_{\mu\nu}F^{\mu\nu} = -\frac{1}{8}\left(F^{(+)}_{\mu\nu}F^{(+)\mu\nu} + F^{(-)}_{\mu\nu}F^{(-)\mu\nu}\right) \tag{4.47a}$$

$$= -\frac{1}{4}\left(\varepsilon_{AB}\varepsilon_{BD}\phi^{AB}\phi^{CD} + \varepsilon_{\dot{A}\dot{C}}\varepsilon_{\dot{B}\dot{D}}\phi^{*\dot{A}\dot{B}}\phi^{*\dot{C}\dot{D}}\right) \tag{4.47b}$$

where ϕ and ϕ^* arc expressed in terms of the potentials ℓ and ℓ^* by (4.45). An alternative way to write the Lagrangian, which would imply (after variation with respect to ϕ and ϕ^*) (4.45), is

$$L = \frac{1}{2}\left[\varepsilon_{AC}\varepsilon_{BD}\phi^{AB}\left(\underset{\sim}{\nabla}^{C\dot{S}}\ell^{D}_{\dot{S}} - \frac{1}{2}\phi^{CD}\right) + \varepsilon_{\dot{A}\dot{C}}\varepsilon_{\dot{B}\dot{D}}\phi^{*\dot{A}\dot{B}}\left(\partial_\mu\ell^{*\dot{D}}_{S}\sigma^{\mu S\dot{C}} - \frac{1}{2}\phi^{*\dot{C}\dot{D}}\right)\right] \quad . \tag{4.48}$$

[Deriving (4.47b) we used the identity

$$Z^\mu W_\mu = \frac{1}{2}\varepsilon_{AB}\overset{*}{\varepsilon}_{\dot{A}\dot{B}}\underset{\approx}{Z}^{A\dot{A}}\underset{\approx}{W}^{B\dot{B}} \tag{4.49}$$

valid for any two 4-vectors Z and W.]

The standard Lagrangian formalism for the electromagnetic field requires the usual sacrifices. Since $\ell^A_{\dot{B}}$ belongs to the representation (1/2, 1/2) of SL(2,C) [and not to a representation of the type (j, j+1) which appears in the above-mentioned Weinberg's theorem] its covariant quantum field theoretic treatment requires an indefinite metric space H. Moreover, Maxwell's equation $\partial_\mu F^{\mu\nu} = 0$ is only satisfied in a (weak) mean value sense, for matrix elements between vectors of a physical subspace H' (see, e.g., [F1]).

4.4 Remark on the Quantization of Higher-Dimensional Conformal Orbits

We saw in the foregoing sections that the quantization of a 0-mass particle system of any helicity amounts essentially to constructing the corresponding unitary irreducible representations of the conformal group SU(2,2). The same should be true for the quantum description of conformal 1-particle systems of varying positive mass. As we saw in Chap.3 the corresponding classical phase spaces are 8, 10 or 12 dimensional depending on whether the particle is spinless, has a fixed positive spin, or its spin can vary in a finite interval. All positive energy unitary irreducible (ray) representations of SU(2,2) have been classified by *Mack* [M1]. What remains to be done is to find the action of the dynamical variables x and $\pi_\pm$ in the corresponding representation spaces [since these variables do not belong to the enveloping algebra of SU(2,2)]. An explicit realization of this quantized picture should provide a useful tool for the study of the quantum dynamics of relativistic interacting particles in the lines of Ref. [T4].

Appendix

A. Clifford Algebra Approach to Twistors. Relation to Dirac Spinors

A.1 Clifford Algebra of O(6,ℂ) and Bitwistor Representation of the Lie Algebra SO(6,ℂ)

The Clifford algebra Cliff X associated with an inner product vector space X can be realized as a matrix algebra with the following property. There is a linear isomorphism $\Gamma: X \to \text{Cliff } X$, such that if $Z, Z' \in X$ then the anticommutator of the matrices ΓZ and $\Gamma Z'$ is twice the inner product (Z,Z'):

$$[\Gamma Z, \Gamma Z']_+ = 2(Z,Z') \tag{A.1}$$

(we omit the unit matrix on the right-hand side).

We shall study the Clifford algebra of the vector space $\mathbb{C}^6$ equipped with an $O(6,\mathbb{C})$ invariant (non-degenerate) inner product. Having in mind applications to the Euclidean (and Minkowski space) conformal group $O(5,1)$ $O(4,2)$ we shall use a basis e_a, $a=1,2,3,4,5,6$ (or $a=0,1,2,3,5,6,$) for which

$$(e_a,e_b) = \eta_{ab} = \text{diag}(++++,\ +-) \quad [\text{or diag}\ (-+++,\ +-)] \quad . \tag{A.2}$$

The corresponding Clifford units $\Gamma_a = \Gamma e_a$ satisfy the anticommutation relations

$$[\Gamma_a,\Gamma_b]_+ = 2\eta_{ab} \quad , \quad a,b = 1,1,3,4,(0),5,6 \quad . \tag{A.3}$$

All (faithful) irreducible representations of this algebra are equivalent and given by 8×8 matrices. [More generally, there is a single 2^ν dimensional irreducible representation of Cliff $(\mathbb{C}^n,\eta)$ for $n=2\nu$ and for $n=2\nu+1$; for various expositions of this fundamental result see [C1],[J1],[B5],[A9],[D2],[H9],[S3].]

If Γ_a satisfy (A.3), then the matrices

$$\Gamma_{ab} = \frac{1}{4}[\Gamma_b,\Gamma_a] \tag{A.4}$$

provide an 8-dimensional (reducible) representation of the mathematical generators of the Lie algebra $SO(6,\mathbb{C})$:

$$[\Gamma_{ab},\Gamma_{cd}] = \eta_{ac}\Gamma_{bd} - \eta_{ad}\Gamma_{bc} + \eta_{bd}\Gamma_{ac} - \eta_{bc}\Gamma_{ad} \quad ; \tag{A.5}$$

it will be called the *bitwistor* representation. Its reducibility can be deduced from the fact that the Casimir operator

$$\Gamma_7 = \frac{1}{3!}\,\varepsilon_{abcdef}\,\Gamma^{ab}\Gamma^{cd}\Gamma^{ef} \quad (= \Gamma^1\Gamma^2\Gamma^3\Gamma^4\Gamma^5\Gamma^6) \tag{A.6}$$

anticommutes with Γ_a,

$$[\Gamma_7,\Gamma_a]_+ = 0 (\Gamma_7^2 = 1) \tag{A.7}$$

(and hence is not a multiple of the unit matrix). The bitwistor representation splits into two inequivalent irreducible 4-dimensional representations: the twistor representation and its dual. They are intertwined by the *space reflection operator*

$V(I_s) = i\Gamma^4\Gamma^5\Gamma^6 (= -i\Gamma_4\Gamma_5\Gamma_6$, satisfying

$$V(I_s)\Gamma_a V^{-1}(I_s) = I_s\Gamma_a = \begin{cases} \Gamma_a & a = 4,5,6 \\ -\Gamma_a & a = 1,2,3 \end{cases} \quad . \tag{A.8}$$

[The factor i corresponds to the convention that twice repeated reflection is a rotation on 2π whose spinor representation is the multiplication by -1, rather than the identity transformation; with the choice (A.8) we have $V^2(I_s) = 1$.]

The generators (A.4) are reduced in the following (call it *Cartan*) basis:

$$\Gamma_a = \begin{pmatrix} 0 & \beta_a \\ \check{\beta}_a & 0 \end{pmatrix} , \quad \text{in which} \quad \Gamma_{ab} = \begin{pmatrix} \gamma_{ab} & 0 \\ 0 & \check{\gamma}_{ab} \end{pmatrix} , \quad \Gamma_7 = \begin{pmatrix} \mathbb{1} & 0 \\ 0 & -\mathbb{1} \end{pmatrix} ,$$

$$V(I_s) = i\begin{pmatrix} 0 & -\beta_4\check{\beta}_5 \\ \beta_4\check{\beta}_5 & 0 \end{pmatrix} . \tag{A.9}$$

Here β_a and $\check{\beta}_a$ are the 4×4 matrices of Sect.1.4, satisfying (1.36), and $\gamma_{ab}, \check{\gamma}_{ab}$ are given by (1.40).

A.2 The Homomorphism SL(4,ℂ)→SO(6,ℂ). Inequivalent 4-Dimensional Analytic Representations of SL(4,ℂ)

The bitwistor representation of SO(6,ℂ) can be integrated to a (single valued, reducible) representation of SL(4,ℂ). In the reduced basis (A.9) it is given by

$$V(\lambda) = \exp\left(\frac{1}{2}\,\lambda^{ab}\Gamma_{ab}\right) = \begin{pmatrix} V(\lambda) & 0 \\ 0 & \check{V}(\lambda) \end{pmatrix} \quad (\lambda^{ab} = -\lambda^{ba} \in \mathbb{C}) \quad , \tag{A.10}$$

where V and $\check{V}$ are expressed in terms of the generators γ_{ab} and $\check{\gamma}_{ab}$ by (2.13b). The irreducible representations V and $\check{V}$ are inequivalent (since they correspond to different eigen-values of the Casimir operator (A.6). All other 4-dimensional irreducible representations of SL(4,ℂ) that are analytic in the parameters λ, are equivalent to one of them. The equivalence between $\check{V}(\lambda)$ and ${}^tV^{-1}(\lambda)$ is realized by the matrix B satisfying ${}^t(B\beta_a) = -B\beta_a$ (1.45). (The passage to complex parameters λ does not alter the straightforward demonstration of this fact already mentioned in Sect. 1.4). A similar calculation shows that in the Euclidean basis

$$\bar{V}(\lambda) \equiv \overline{V(\bar{\lambda})} = BV(\lambda)B^{-1} \quad , \quad V(\bar{\lambda})^{-1} = V(\lambda)[=B^{t}V(\lambda)^{-1}B^{-1}] \tag{A.11}$$

(in infinitesimal form $\bar{\gamma}_{ab} = B\gamma_{ab}B^{-1}$ and $\gamma^{*}_{ab} + \gamma_{ab} = 0$, for a,b, =1,2,3,4,5,6).

There is a two to one homomorphism between the bitwistor representation (A.10) of $SL(4,\mathbb{C})$ and the defining (6-vector) representation of the complex orthogonal group $SO(6,\mathbb{C})$ such that

$$V(\lambda)\Gamma Z V(\lambda)^{-1} = \Gamma\Lambda(\lambda)Z \quad \text{or} \quad V\beta Z\check{V}^{-1} = \beta\Lambda(\lambda)Z, \ \check{V}\beta Z V^{-1} = \check{\beta}\Lambda Z \ . \tag{A.12}$$

Using the orthonormality relation

$$\mathrm{tr}\Gamma_a\Gamma_b = 8\eta_{ab} \ , \tag{A.13}$$

we find

$$\Lambda(\lambda)^a_b = \frac{1}{8}\,\mathrm{tr}\left[\Gamma^a V(\lambda)\Gamma_b V(\lambda)^{-1}\right] = \frac{1}{4}\,\mathrm{tr}(\beta^a V\beta_b V^{-1})$$

$$= \delta^a_b + \frac{1}{2}\lambda^{cd}(\eta_{bc}\delta^a_d - \eta_{bd}\delta^a_c) + O(\lambda^2) \ . \tag{A.14}$$

The space $T \oplus \tilde{T}$ $(=\mathbb{C}^8)$ of bitwistors admits an $SL(4,\mathbb{C})I_s$-invariant skewsymmetric form $\chi C \phi = -\phi C \chi (\chi,\phi \in T \oplus \tilde{T})$ where the antisymmetric matrix C satisfies

$$^{t}\Gamma_a C = C\Gamma_a \quad , \quad ^{t}\Gamma_7 C = -\Gamma_7 C \ . \tag{A.15}$$

In the basis (A.9) a C with this property is

$$C = \begin{pmatrix} 0 & -^{t}B \\ B & 0 \end{pmatrix} \quad (= -^{t}C) \tag{A.16}$$

(it is determined up to a phase factor by (A.15) and by the normalization condition $\det C = 1$). The invariance condition

$$^{t}V(\lambda)CV(\lambda) = C \quad , \quad \text{or} \quad ^{t}\Gamma_{ab}C + C\Gamma_{ab} = 0 \tag{A.17a}$$

$$^{t}V(I_s)CV(I_s) = C \tag{A.17b}$$

is a straightforward consequence of (A.15). [Note the role of the factor i in (A.8) for the validity of (A.17b)].

Exercise A1. Find the maximal connected Lie group G of linear transformations in $\mathbb{C}^8$ that leaves the bilinear form $\chi C\phi$ invariant.

[*Answer*: G is the 36-parameter complex symplectic group $Sp(4,\mathbb{C})$ (in 8 space) generated by $\Gamma_{ab}, \Gamma_{abc} = \frac{1}{3}(\Gamma_{ab}\Gamma_c + \Gamma_{bc}\Gamma_a + \Gamma_{ca}\Gamma_b)$, and $\frac{1}{2}\Gamma_7$.

There exists also a symmetric bilinear form

$$\chi B\phi \quad B = C\Gamma_7 \tag{A.18}$$

invariant under proper $SL(4,\mathbb{C})$ transformations $B\Gamma_{ab} + {}^t\Gamma_{ab}B = 0$). [It changes sign under space reflection, since $V(I_s)\Gamma_7 V(I_s)^{-1} = -\Gamma_7$.]

Exercise A2. Find the maximal connected Lie group G_B of linear transformations in $\mathbb{C}^8$ that leaves the form (A.18) invariant.

[*Answer*: G_B is the 28-parameter complex orthogonal group $SO(8,\mathbb{C})$ generated by $\frac{1}{2}\Gamma_a, \Gamma_{ab}, \frac{1}{2}\Gamma_a\Gamma_7, \frac{1}{2}\Gamma_7$.]

The restriction to the Minkowskian real form $SU(2,2)$ of $SL(4,\mathbb{C})$ leads to the existence of a pair of invariant Hermitian forms in $\mathbb{T}\oplus\tilde{\mathbb{T}}$:

$$\bar{\phi}A\chi \quad \text{and} \quad \bar{\phi}A\Gamma_7\chi \tag{A.19a}$$

where A satisfies ($\bar{A} = A$ and)

$$A\Gamma_a = \Gamma_a^* A \quad \text{for} \quad a = 0,1,2,3,5,6 \quad (\text{and } 7) \quad ; \tag{A.19b}$$

in a basis in which $\Gamma_a^* = \eta_{aa}\Gamma_a$ [in particular, in the basis (A.9)] we can set

$$A = \Gamma_6\Gamma_4 = i\Gamma_6\Gamma_0 \left[= \begin{pmatrix} -A & 0 \\ 0 & A \end{pmatrix} \quad \text{for} \quad A = i\beta^0 = -\beta_e \right] \tag{A.19c}$$

[A satisfying (1.42)].

Exercise A3. Verify the relations

$$V(I_s)^* A V(I_s) = A \quad , \quad V(I_s)^* A\Gamma_7 V(I_s) = -A\Gamma_7 \quad . \tag{A.20}$$

Find the maximal (connected) subgroup of $SL(8,\mathbb{C})$ that leaves both sesquilinear forms (A.19) invariant.

[*Answer*: It is the 31-parameter "chiral" group $S[U(2,2)\times U(2,2)]$ generated by Γ_{ab}, $\Gamma_7\Gamma_{ab}$ and $\frac{i}{2}\Gamma_7$, $a,b = 0,1,2,3,5,6$.]

The covering Spin (5,1) of the Euclidean conformal group is another real form of $SL(4,\mathbb{C})$ [the bitwistor representation of the (real) Lie algebra Spin (5,1) is generated by Γ_{ab} for $a,b = 1,2,3,4,5,6$]. Equations (A.10 and 11) imply that there are two Spin (5,1) invariant Hermitian forms in C^8: the scalar form $i\bar{\phi}\Gamma_6\chi$ and the pseudoscalar form $\bar{\phi}\Gamma_7\chi_6$.

Although the representations V and $\bar{V}$ of Spin(5,1) are equivalent [according to (A.11)] there are no real Euclidean twistors (the only solution of the Spin(5,1) invariant reality condition $B^{-1}\bar{\zeta} = \zeta$ being $\zeta = 0$, see [M9]). There are, however, self-conjugate bivectors $\underset{1}{\zeta} \wedge \underset{2}{\zeta} (= B^{-1} \otimes B^{-1} \underset{1}{\bar{\zeta}} \wedge \underset{2}{\bar{\zeta}})$, corresponding to points of the compactified Euclidean space S^4. They have the form $\zeta \wedge \bar{B}\bar{\zeta}$.

Exercise A4. Show that Spin(5,1) is isomorphic to the group SL(2,H) of 2×2 matrices with quaternionic elements satisfying $\det\begin{pmatrix} a & b \\ c & d \end{pmatrix} = \det(a - b\, d^{-1}c)\det d = a^*ad^*d + b^*bc^*c - a^*bd^*c - b^*ac^*d = 1$.

A.3 Conformal Dirac Spinors

Minkowski space twistors do not coincide with Dirac spinors although they have the same Lorentz [or rather $SL(2,\mathbb{C})$] transformation properties. That is already clear from the fact that a twistor goes under the representation (A.8) of space reflection I_s into a dual twistor while for (4-component) Dirac spinors the corresponding representation of I_s does not lead out of the space. Dirac spinors can be obtained from 8-component bitwistors by projecting on an eigensubspace of the dilation generator $\Gamma_{65} = \frac{1}{2}\Gamma_5\Gamma_6$.

We shall describe two alternative ways to imbed a Dirac spinor into an 8-component bitwistor, and conversely to project a Dirac spinor out of an 8-spinor (see also [F5]).

We start with a general 4-component spinor field $\psi(x)$ of dimension d, and will then exhibit the peculiarities of a canonical (massless) Dirac field obtained for $d = 3/2$.

(i) The first approach, which goes back to *Dirac* ([D1]; see also [M2][M11,12]), associates with each $\psi(x)$ an 8-component spinor field $\Psi(\xi)$ of homogeneity degree $-(d+1/2)$. The x-space counterpart of $\Psi(\xi)$

$$k^{d+1/2}T(-x)\Psi(\xi) \quad , \tag{A.21}$$

where $T(a)$ is the bitwistor representation of translations, is transforming under a non-decomposable representation of the conformal group that splits into the direct sum of two 4-component spinors $\psi_-(x)[=\psi(x)]$ of dimension d and $\psi_+(x)$ of dimension $d+1$ when restricted to the Weyl subgroup Aut P of SU(2,2) (generated by Poincaré transformations and dilations). The original spinor field $\psi(x)[=\psi_-(x)]$ spans a conformal invariant subspace whose complement, $\psi_+(x)$, is not invariant under special conformal transformations; instead, we have

$$\frac{1}{i}[\psi_+(x),K_\mu] = \left\{2x_\mu(x\partial + d + 1) - x^2\partial_\mu + \frac{1}{2}[\gamma_\mu,x\gamma]\right\}\psi_+(x) + i\gamma_\mu\psi_-(x) \quad . \tag{A.22}$$

In order to exhibit this Aut P-content of $\Psi(\xi)$ it is convenient to use another (call it *Dirac*) *realization* of Γ_a in which the eigensubspaces of the 8-dimensional dilation generator are singled out:

$$2\Gamma^D_{65} = \Gamma^D_5\Gamma^D_6 = -\tau_3 \otimes \mathbb{1} = \begin{pmatrix} -1 & 0 \\ 0 & \mathbb{1} \end{pmatrix} \quad . \tag{A.23}$$

The following direct product realization fits this requirement

$$\Gamma^D_\mu = \tau_3 \otimes \gamma_\mu \quad , \quad \Gamma^D_5 = \tau_2 \otimes \mathbb{1} \quad , \quad \Gamma^D_6 = -i\tau_1 \otimes \mathbb{1} \quad , \tag{A.24}$$

$$\Gamma^D_7 = \tau_3 \otimes \gamma_5 \quad ; \quad V^D(I_s) = \mathbb{1} \otimes \gamma^0 \quad ;$$

$$A^D = -\tau_2 \otimes A = i\tau_2 \otimes \gamma_0 \quad . \tag{A.25}$$

Exercise A5. Setting in the basis (A.24,25)

$$k^{d+1/2}T^D(-x)\Psi(\zeta) = \begin{pmatrix}\psi_+ & (x)\\ \psi_- & (x)\end{pmatrix} \tag{A.26}$$

demonstrate that the manifestly invariant "Dirac equation"

$$\Gamma\xi\Gamma \frac{\partial}{\partial\xi}\Psi(\xi) = 0 \tag{A.27}$$

is equivalent to

$$(3 - 2d)\psi_+(x) + i\gamma\partial\psi_-(x) = 0 \tag{A.28}$$

which reduces to the usual Dirac equation for canonical dimension $d = 3/2$.

[*Hint*: Note that in the above basis

$$T^D(-x)\Gamma^D\xi\Gamma^D \frac{\partial}{\partial\xi} T^D(x) = \frac{1}{2}\left(\Gamma_5^D + \Gamma_6^D\right)\Gamma^{\mu D}\partial_\mu$$

$$= \begin{pmatrix}0 & i\gamma\partial\\ 0 & 0\end{pmatrix} \quad . \tag{A.29}$$

Exercise A6. Let $\beta_\mu = \gamma_\mu$; find a 2×2 block-matrix S with 4×4 matrix entries such that $\Gamma_a = S\Gamma_a^D S^{-1}$, $\det S = 1$.

[*Answer*:

$$S = e^{-i(\pi/4)}\begin{pmatrix} i\,\frac{1+\gamma_5}{2} & \frac{\gamma_5-1}{2}\\ i\,\frac{1-\gamma_5}{2} & \frac{\gamma_5+1}{2}\end{pmatrix} = S^{*-1} \quad .$$

(ii) The second possibility (used in [M4,T5]) consists in extending $\psi(x)$ to an 8-component spinor field $\Psi(\xi)$ of homogeneity degree $-(d-1/2)$ that satisfies the subsidiary condition

$$\Gamma\xi\Psi(\xi) = 0 \quad . \tag{A.30}$$

Exercise A7. Demonstrate that if $\Psi(\xi)$ satisfies (A.30) [as well as the homogeneity condition $\rho^{(d-1/2)}\Psi(\rho\xi) = \Psi(\xi)$], then

$$\Psi(x) \equiv k^{d-1/2}T(-x)\Psi(\xi) = \frac{1}{2}(1 + \Gamma_6\Gamma_5)\Psi(x) \quad . \tag{A.31}$$

Exercise A8. Show that

$$\Pi_\xi \equiv \Gamma\xi\,\frac{\Gamma_5 - \Gamma_6}{2k} \tag{A.32}$$

is a projection operator of rank 4 (i.e. $\Pi_\xi^2 = \Pi_\xi$, $\mathrm{tr}\Pi_\xi = 4$). Demonstrate that the subsidiary condition (A.30) is equivalent to

$$(\Pi_\xi - 1)\Psi(\xi) = 0 \quad . \tag{A.33}$$

The free (0-mass) Dirac equation (for $d = 3/2$) assumes a somewhat awkward form

$$\Gamma\xi\Gamma \frac{\partial}{\partial\xi} \frac{\Gamma_5 - \Gamma_6}{2} \Psi(\xi) = 0 \tag{A.34}$$

in the above manifestly covariant picture.

We observe finally that the two pictures are related by the (non-invertible) mapping

$$\Psi^{(ii)}(\xi) = \Gamma\xi\Psi^{(i)}(\xi) \quad . \tag{A.35}$$

References

[A1] R. Abłamowicz, J. Mozrzymas, Z. Oziewicz, J. Rzewuski: Spinor space. Rep. Math. Phys. **14**, 89-100 (1978)

[A2] R. Abłamowicz, Z. Oziewicz, R. Rzewuski: On the projection of the spinor space on the Minkowski space. Bull. Acad. Polon. Sci. Ser. Sci. Phys. Astronom. **27**, 201-203 (1980)

[A3] L.P. Hughston, R.S. Ward (eds.): *Advances in Twistor Theory*, Research Notes in Mathematics **37** (Pitman, London 1979)

[A4] P.C. Aichelburg: Curvature collineations for gravitational pp waves. J. Math. Phys. **11**, 2458-1462 (1970)

[A5] E. Angelopoulos, F. Bayen, M. Flato: On the localizability of massless particles. Phys. Scr. **9**, 173-183 (1974);
E. Angelopoulos, M. Flato: On unitary implementability of conformal transformations. Math. Phys. Lett. **2**, 405-412 (1978)

[A6] E. Angelopoulos, M. Flato, C. Fronsdal, D. Sternheimer: Massless particles, conformal group and de Sitter universe. Phys. Rev. D**23**, 1278-1289 (1981)

[A7] M.F. Atiyah: *Geometry of Yang-Mills Fields* (Lezione Fermiane, Pisa, Scuola Normale Superiore 1979)

[A8] M.F. Atiyah, R. Bott: Yang-Mills and bundles over algebraic curves. Proc. Indian Acad. Sci. Math. Sci. **90** (1), 11-20 (1981)

[A9] M.F. Atiyah, R. Bott, A. Shapiro: Clifford models. Topology **3**, Suppl. 1, 3-38 (1964)

[A10] M.F. Atiyah, V.G. Drinfeld, N.J. Hitchin, Yu.I. Manin: Construction of instantons. Phys. Lett. **65**A, 185-187 (1978)

[A11] M.F. Atiyah, N.J. Hitchin, I.M. Singer: Self-duality in four-dimensional Riemannian geometry. Proc. Roy. Soc. A**362**, 425-461 (1978)

[A12] M.F. Atiyah, R.S. Ward: Instantons and algebraic geometry. Commun. Math. Phys. **55**, 117-124 (1977)

[B1] F. Bayen, M. Flato, C. Fronsdal, A. Lichnerowicz, D. Sternheimer: Deformation theory and quantization, I. Deformation of symplectic structures, II. Physical applications. Ann. Phys. N.Y. **111**, 61-110, 111-151 (1978)

[B2] F. Bayern, J. Niederle: Localizability of massless particles in the framework of the conformal group. Preprint IC/81/115, Trieste (1981)

[B3] A.A. Beilinson, S.I. Gel'fand, Yu.I. Manin: An instanton is determined by its complex singularities. Funk. Anal. Prilozh. **14**, 48-49 (1980) [English transl.: Func. Anal. Appl. **14**, 118-119 (1980)]

[B4] F.A. Berezin: Quantization, Izv. Akad. Nauk SRR **38**, 1116-1175 (1974 [in Russian]

[B5] R. Brauer, H. Weyl: Spinors in n-dimensional space. Am. J. Math. **57**, 425 (1935)

[B6] P. Budinich: Quarks as conformal semispinors. ICTP Preprint IC/79/88, Trieste (1979); Reflection groups and internal symmetry algebras. Nuovo Cim. **53**A, 31-81 (1979); On conformal covariance of spinor field equations, ICTP Preprint IC/79/164, Trieste (1979); On "conformal spinor geometry": an attempt to "understand" internal symmetry. ICTP Preprint IC/81/189, Trieste (1981)

[B7] P. Budinich: Pure spinors and quadric Grassmannians. ISAS Int. Rep. 37/85/EP; Pure spinors for conformal extensions of space-time, ISAS Int. Rep. 1/86/EP, Trieste

[B8] P. Budinich, P. Furlan: On a "conformal spinor field equation". Phys. Lett. **107**B, 434-436 (1981)

[B9] P. Budinich, P. Furlan, R. Raczka: Weyl and conformal invariant field theories. Nuovo Cimento **52**A, 191-246 (1979)
[B10] P. Budinich, A. Trautman: Remarks on pure spinors. ISAS Preprint 87/85/EP, Trieste
[C1] E. Cartan: *Leçons sur la théorie des spineurs* (Paris, Hermann 1937) [English transl.: *The Theory of Spinors*; with a forward by R. Streater (Paris, Hermann 1966 p.157]
[D1] P.A.M. Dirac: Wave equations in conformal space. Ann. Math. **37**, 429-442 (1936); Relativistic wave equations. Proc. Roy. Soc. London Ser. A **155**, 447-459 (1936)
[D2] P.A.M. Dirac: *Spinors in Hilbert Space* (Plenum, New York 1974)
[D3] Y. Dothan, M. Gell-Mann, Y. Ne'eman: Series of hadron energy levels as representations of non-compact groups. Phys. Lett. **17**, 148-151 (1965)
[D4] A. Douady: La transformation de Penrose. Séminaire Ecole Normale Supérieure 1977-1978, Exp. N.3,4
[D5] B.A. Dubrovin, S.P. Novikov, A.T. Fomenko: *Modern Geometry, Methods and Applications* (Nauka, Moscow 1979)
[E1] M.G. Eastwood: On the twistor description of massive fields. Proc. Roy. Soc. London A**374**, 431-445 (1981)
[E2] M.G. Eastwood: Twistor theory (the Penrose transform). Lectures presented at the Summer Seminar on Complex Analysis SMR/73-16, Trieste (1980)
[E3] M.G. Eastwood, M.L. Ginsberg: Duality in twistor theory. Preprint Oxford (1980)
[E4] M.G. Eastwood, R. Penrose, R.O. Wells, Jr.: Cohomology and massless fields. Commun. Math. Phys. **78**, 305-351 (1981)
[E5] L.P. Eisenhart: *Riemannian Geometry* (Princeton Univ. Press 1964)
[F1] R. Ferrari, L. Picasso, F. Strocchi: Some remarks on local operators in quantum electrodynamics. Commun. Math. Phys. **35**, 25-38 (1974)
[F2] A. Ferber: Supertwistors and conformal supersymmetry. Nucl. Phys. B**123**, 55-64 (1978)
[F3] M. Flato, C. Fronsdal: Quantum field theory of singletons. J. Math. Phys. **22**, 1100-1105 (1981)
[F4] M. Flato, D. Sternheimer: Remarques sur les automorphismes causals de l'espace temps. C.R. Acad. Sci. Paris **263**A, 935-936 (1966)
[F5] P. Furlan: Are SO(2N,2)-covariant spinor wave equations also "conformal"-covariant? Nuovo Cimento **71**A, 43-71 (1982)
[G1] S.G. Gindikin, G.M. Khenkin: Integral geometry for $\bar{\partial}$-cohomology in q-linear concave domains in $\mathbb{CP}^n$. Funk. Anal. Prilozh. **12**, N.4, 6-23 (1978) [English transl.: Func. Anal. Appl. **12**, 247-261 (1979)]
[G2] S.G. Gindikin, G.M. Khenkin: Complex integral geometry and Penrose representations for solutions of Maxwell equations. Teor. Mat. Fiz. **43**, 18-31 (1980); see also: Penrose transform and complex integral geometry, in *Sovremennye Problemy Matematiki*, Vol.17 (VINITI, Moscow 1981) [in Russian]
[G3] P.S. Green, J. Isenberg, P. Yasskin: Non-self-dual gauge fields. Phys. Lett. **78**B, 462-464 (1978)
[G4] F. Gürsey, H.C. Tze: Complex and quaternionic analyticity in chiral and gauge theories. Ann. Phys. NY **128**, 29-130 (1980)
[H1] R.O. Hanson, E.T. Newman: A complex Minkowski space approach to twistors. Gen. Relativ. Gravit. **6**, 361-385 (1975)
[H2] R.O. Hanson, E.T. Newman, R. Penrose, K.P. Tod: The metric and curvature properties of H space. Proc. Roy. Soc. A**363**, 445-468 (1978)
[H3] G.M. Henkin, Yu.I. Manin: Twistor description of classical Yang-Mills-Dirac fields. Phys. Letters **95**B, 405-408 (1980)
[H4] W.A. Hepner: The inhomogeneous Lorentz group and the conformal group. Nuovo Cimento **26**, 351-358 (1962)
[H5] N.J. Hitchin: Polygons and gravitons. Math. Proc. Comb. Philos. Soc. **85**, 465-476 (1979)
[H6] N.J. Hitchin: Linear field equations on self-dual spaces. Proc. Roy. Soc. A**370**, 173-191 (1980)
[H7] L.P. Hughston: *Twistors and Particles*, Lecture Notes in Physics, Vol. 97 (Springer, Berlin Heidelberg 1979)
[H8] L.P. Hughston: The Twistor particle programme, Preprint KFKI-1980-18, Budapest (1980)

[H9] D. Husemoller: *Fibre Bundles*, 2nd ed. (Springer, New York 1975) Chap.11, Sect.4 "Clifford Algebra of a Quadratic Form", pp.144-146 and Sect.5 "Calculations of Clifford Algebras", pp.146-149
[I1] J. Isenberg, Ph.B. Yasskin: "Twistor Description o- Non-Self-Dual Yang-Mills Fields, in *Complex Manifold Techniques in Theoretical Physics*, ed. by D. Lerner, P. Sommers (Pitman, London 1979);
J. Isenberg, Ph.B. Yasskin, P.S. Green: Non-self dual Yang-Mills fields. Phys. Lett. B**78**, 462-467 (1982)
[J1] P.Jordan, E. Wigner: Über das Paulische Äquivalenzverbot. Z. Phys. **47**, 631-658 (1928) (See, in particular, Zusatz bei der Korrektur, pp.650-651)
[K1] A.A. Kirillov: *Elements of the Theory of Representation* (Springer, Berlin, Heidelberg 1976)
[K2] M. Ko, M. Ludvigsen, E.T. Newman, K.P. Tod: The theory of H space. Phys. Rep. **71**, 51-139
[K3] M. Ko, E.T. Newman, R. Penrose: The Kähler structure of asymptotic twistor space. J. Math. Phys. **18**, 58-64 (1977)
[K4] S. Kobayashi, K. Nomizu: *Foundations of Differential Geometry* (Interscience, New York 1963), Vol.II (1969)
[K5] M. Kuiper: On conformal flat spaces in the large. Ann. Math. **50**, 916-924 (1949)
[K6] B. Kurşunoglu: *Modern Quantum Theory* (Freeman, San Francisco 1962) p.257
[K7] W. Kopczynski, L.S. Woronowicz: A geometrical approach to the twistor formalism. Rep. Math. Phys. **2**, 35-51 (1971)
[L1] D. Lerner: The inverse twistor function for positive frequency fields. Twistor Newsletter, N.5 (Oxford, July 1977)
[L2] J. Levine: Groups of motion of conformally flat spaces, I and II. Bull. Am. Math. Soc. **42**, 418-422 (1936) and **45**, 766-773 (1939)
[L3] W. Lisiecki, A. Odzijewicz: Twistor flag spaces as phase spaces of conformal particles. Lett. Math. Phys. **3**, 325-334 (1979)
[L4] M. Ludvigsen, E.T. Newman, K.P. Tod: Asymptotically flat H spaces. J. Math. Phys. **22**, 818-823 (1981)
[M1] G. Mack: All unitary ray representations of the conformal group SU(2,2) with positive energy. Commun. Math. Phys. **55**, 1-28 (1977)
[M2] G. Mack, Abdus Salam: Finite component field representations of the conformal group. Ann. Phys. NY **53**, 174-202 (1969)
[M3] G. Mack, I.T. Todorov: Irreducibility of the ladder representations of U(2,2) when restricted to the Poincaré subgroup. J. Math. Phys. **10**, 2078-2085 (1969)
[M4] G. Mack, I.T. Todorov: Conformal invariant Green functions .ithout ultraviolet divergences. Phys. Rev. D**8**, 1764-1787 (1973)
[M5] D. Maison: Some facts about classical non-linear σ-models. Ma -Planck Institute Preprint MPI-PA E/PTh 52/79 (November 1979)
[M6] Yu.I. Manin: Gauge fields and holomorphic geometry. Sovrem. Probl. Mat. **17** (Moscow, VINITI, 1981) [in Russian]
[M7] R. Marnelius: Manifestly conformally covariant description of spinning and charged particles. Phys. Rev. D**20**, 2091-2095 (1979)
[M8] R. Marnelius, B. Nilsson: Manifestly conformally covariant field equations and a geometrical understanding of mass. Phys. Rev. D**22**, 830-838 (1980)
[M9] L. Michel, L. Radicati: The geometry of the octet. Ann. Inst. H. Poincaré **18**, 185-214 (1973)
[M10] V. Molotkov, S. Petrova: Scalar representations of conformal superalgebra. JINR Report E2-10126, Dubna (1976); Unitary scalar representations of the conformal superalgebra, JINR Report P2-10127, Dubna (1976) [in Russian]
[M11] Y. Murai: On the group of transformations in six-dimensional space. Prog. Theor. Phys. **9**, 147-168 (1953); **11**, 441-448 (1954)
[M12] Y. Murai: New wave equations for elementary particles. Nucl. Phys. **6**, 489-503 (1958)
[N1] E.T. Newman, R. Penrose: New conservation laws for zero-rest-ass fields in asymptotically flat space-times. Proc. Roy. Soc. **305**A, 175-204 (1968)
[N2] E.T. Newman, K.P. Tod: A note on left-flat space times. J. Math. Phys. **21**, 874-877 (1980)
[O1] A. Odzijewicz: A model of conformal kinematics. Int. J. Theor. Phys. **15**, 375-593 (1976)

[O2] A. Odzijewicz: Conformal invariant symplectic structures (semi-simple case). Rep. Math. Phys. **12**, 407-421 (1977)
[P0] R. Penrose: "Conformal Treatment of Infinity", in *Relativity, Groups and Topology*, ed. by C.M. De Witt, B. de Witt, Les Houches Summer School, 1963 (Gordon and Breach, New York 1964) pp.565-584
[P1] R. Penrose: Twistor algebra. J. Math. Phys. **8**, 345-366 (1967); R. Penrose, W. Rindler: *Spinors and Space-Time*, Vol.1. Two-spinor calculus and relativistic fields (Cambridge Univ. Press, 1984)
[P2] R. Penrose: Twistor quantization and curved space time. Int. J. Theor. Phys. **1**, 61-99 (1969)
[P3] R. Penrose: Solutions of zero-mass equations. J. Math. Phys. **10**, 38-39 (1969)
[P4] R. Penrose: "Twistor theory, Its Aims and Achievements", in *Quantum Gravity: An Oxford Symposium*, ed. by C.J. Isham, R. Penrose, D.W. Sciama (Clarendon Press, Oxford 1975) pp.268-407
[P5] R. Penrose: "Twistors and Particles – an Outline", in *Quantum Theory and the Structure of Time and Space*, ed. by L. Castel, M. Drieschner, C.F. von Weizsäcker (Carl Kanser Verlag, München 1975) pp.129-145
[P6] R. Penrose: The twistor program. Rep. Math. Phys. **12**, 65-76 (1977); Is nature complex? In *The Encyclopedia of Ignorance* (1979)
[P7] R. Penrose: Nonlinear gravitons and curved twistor theory. Gen. Relativ. Gravit. **7**, 31-52 (1976)
[P8] R. Penrose: Massless fields and sheaf cohomology. Twistor Newsletter N.5 (Oxford, July 1977); "On the Twistor Description of Massless Fields", in *Complex Manifold Techniques in Theoretical Physics*, ed. by D.E. Lerner, P.D. Sommers, Research Notes in Mathematics, Vol.32 (Pitman, London 1979) pp.55-91
[P9] R. Penrose, M.A.H. MacCallum: Twistor theory: an approach to the quantisation of fields and space-time. Phys. Rep. C**6**, 241-361 (1972)
[P10] R. Penrose, G.A.J. Sparling, S.T. Tsun: Extended Regge trajectories. J. Phys. A: Math. Gen. **11**, No.9, L231-L235 (1978)
[P11] R. Penrose, R.S. Ward: "Twistors for Flat and Curved Space-Time", in *Einstein Centennial Volume*, ed. by P.G. Bergmann, J.N. Goldberg, A.P. Held (1979)
[P12] Z. Perjés: Twistor variables in relativistic mechanics. Phys. Rev. D**11**, 2031-2041 (1975)
[P13] Z. Perjés: Perspectives of Penrose theory in particle physics. Rep. Math. Phys. **12**, 193-211 (1977)
[P14] Z. Perjés: Unitary space of particle internal states. Phys. Rev. D**20**, 1857-1876 (1979)
[P15] A.Z. Petrov: *Einstein Spaces* (Pergamon Press, Oxford 1969); see, in particular, Chap.6
[P16] M.H. Pryce: The mass-centre in the restricted theory of relativity and its connection with the quantum theory of elementary particles. Proc. Roy. Soc. A**195**, 62-81 (1984)
[R1] I. Robinson: Null electromagnetic field. J. Math. Phys. **2**, 290-291 (1961)
[R2] J. Rzewuski: Propagators in spinor space. Bull. Acad. Polon. Sci. Ser. Sci. Phys. Astronom. **27**, 201-203 (1980)
[S1] I.E. Segal: Causally oriented manifolds and groups. Bull. Am. Math. Soc. **77**, 958-959 (1971); Interacting quantum fields and the chronometric principle. Proc. Nat. Acad. Sci. USA **73**, 3355-3359 (1976)
[S2] G.A.J. Sparling: "Homology and Twistor Theory", in *Quantum Gravity: An Oxford Symposium*, ed. by C.J. Isham, R. Penrose, D.W. Sciama (Clarendon Press, Oxford 1975)
[S3] Tej Srivastava: On the representation of generalised Dirac (Clifford) algebras. Acta Phys. Austriaca **54**, 287-292 (1982)
[T0] W.E. Thirring: *A Course in Mathematical Physics 2. Classical Field Theory* (Springer New York 1979) see, in particular, pp.151-165
[T1] K.P. Tod: Some symplectic forms arising in twistor theory. Rep. Math. Phys. **11**, 339-346 (1977)
[T2] A.N. Todorov: Applications of the Kähler-Einstein-Calabi-Yao metric to moduli of K3 surfaces. Inventiones Matematicae **61**, 251-265 (1980)

[T3] I.T. Todorov: "Differential Geometric Methods in Relativistic Particle Dynamics. Single Particle Systems", Lectures presented in the Fakultät für Physik, Universität Bielefeld (Spring 1981); see also "Constraint Hamiltonian approach to relativistic particle dynamics", Part I, S.I.S.S.A. Trieste (1980)

[T4] I.T. Todorov: Constraint Hamiltonian mechanics of directly interacting relativistic particles. Universität Bielefeld Preprint BI-TP 81/24 (1981)

[T5] I.T. Todorov, M.C. Mintchev, V.B. Petkova: *Conformal Invariance in Quantum Field Theory* (Pisa, Scuola Normale Superiore, 1978);

[T6] I.T. Todorov: Local field representations of the conformal group and their applications. ZiF Preprint, Bielefeld (1984); Mathematics + Physics, Lectures on Recent Results, Vol.1, L. Streit (ed.) (World Scientific, Singapore 1985) pp.195-338

[U1] A. Uhlmann: Remarks on the future tube. Acta Phys. Pol. **24**, 193 (1963) and Preprint KMU-HEP 7209, Leipzig (1972)

[U2] A. Uhlmann: The closure of Minkowski space. Acta Phys. Pol. **24**, 295-296 (1963)

[W1] R. Ward: The twisted photon. Twistor Newsletter N. 1 (Oxford, March 1976)

[W2] R. Ward: On self-dual gauge fields. Phys. Lett. A**61**, 81-82 (1977)

[W3] R.S. Ward: Ansätze for self-dual Yang-Mills fields. Commun. Math. Phys. **80**, 563-574 (1981)

[W4] S. Weinberg: Feynman rules for any spin, II. Massless particles. Phys. Rev. **134**B, 882-896 (1964); Photons and gravitons in perturbation theory. Derivation of Maxwell's and Einstein's equations. Phys. Rev. **138**B, 988-1002 (1965)

[W5] R.O. Wells, Jr.: Complex manifolds and mathematical physics. Bull. Am. Math. Soc. (New Series) 1, 296-336 (1979); *Complex Geometry in Mathematical Physics*, Notes by R. Pool (Les Presses d l'Université de Montréal, 1982)

[W6] R.O. Wells, Jr.: "Cohomology and the Penrose transform", in *Complex Manifold Techniques in Theoretical Physics*, ed. by D.E. Lerner, P.D. Sommers, Res. Notes Math. **32** (Pitman, London 1979) pp.92-114

[W7] R.O. Wells, Jr.: Hyperfunction solutions of the zero-rest-mass field equations. Commun. Math. Phys. **78**, 567-600 (1981)

[W8] A.S. Wightman: On the localizability of quantum mechanical systems. Rev. Mod. Phys. **34**, 845-872 (1962)

[W9] E. Witten: An interpretation of classical Yang-Mills theory. Phys. Lett. **77**B, 394-398 (1978); Twistor-like transform in ten dimensions. Princeton University Preprint (May 1985)

[W10] N.M.J. Woodhouse: "Twistor Theory and Geometric Quantization", in *Group Theoretical Methods in Physics*, Lecture Notes in Physics, Vol.50 (Springer, Berlin Heidelberg 1976)

[W11] N.M.J. Woodhouse: Twistor cohomology without sheaves. Twistor Newsletter N.2 (Oxford, June 1976)

Subject Index